SAND

The publisher gratefully acknowledges the generous contribution
toward the publication of this book provided by the
Director's Circle of the University of California Press Foundation,
whose members are:

John M. & Jola Anderson

Jacqueline Avant

Nancy & Roger Boas

Beverly Bouwsma

Jamie & Philip Bowles

Robert & Alice Bridges Foundation

Earl & June Cheit

Lloyd Cotsen

Liza Dalby

Sam Davis

Sukey & Gilbert Garcetti

Jean E. Gold

Daniel Heartz

Betty Hine

D. Kern Holoman

David & Sheila Littlejohn

Michael McCone

Thomas & Barbara Metcalf

Jack & Jacqueline Miles

Lucinda Reinold

Roth Family Foundation

Ruth A. Solie

Tania Stepanian

Barry & Marjorie Traub

Patricia Trenton

Stanley & Dorothy Wolpert

MICHAEL WELLAND

SAND

THE NEVER-ENDING STORY

UNIVERSITY OF CALIFORNIA PRESS   Berkeley   Los Angeles

University of California Press
Berkeley and Los Angeles, California

Library of Congress Cataloging-in-Publication Data

Welland, Michael, 1946–.
    Sand : the never-ending story / Michael Welland.
        p.   cm.
    Includes bibliographical references and index.
    ISBN 978-0-520-25437-4 (cloth : alk. paper)
    1. Sand—Miscellanea.   2. Sandstone—Miscellanea.   I. Title.
QE471.15.S25W455   2009
553.6'22—dc22   2008009084

For my parents

CONTENTS

ACKNOWLEDGMENTS

Writing a book of this scope that includes topics with which, on setting out, I had only limited familiarity required a great deal of help. I was fortunate enough to find that help arising from both expected and unexpected quarters—one of the great pleasures has been encountering new friends along the way. Among all those who have supported me in so many different ways, I extend my thanks to the following in particular.

First and foremost, those whom I like to think of as the patriarchs and guides who set the scene for this book: Alan Smith and Chet Bonar (who both, in addition, helped make it a reality), Ray Siever, Aden Womersley, Peter Friend, and Henry "to measure is to know" Weisheit.

For ideas, inspiration, and encouragement: Stephen Bagnold, Marco Bonifazi, Bernie Clow, Christopher Coleman, Jay Critchley, Jim Denevan, Bob Giegengack, Loes Modderman, Larry Nelson, Pete Newman, Bruce Shapiro, and Rob Twigger. All the friends, particularly Laurie and Libby Weisheit, who have assiduously collected sands from exotic locations on my behalf. Terry Stancill, who spontaneously and enthusiastically provided my daughter and me with a guided tour of his Maryland sand operations. The people who, from the start, helped me believe that this book was even possible: Simon Pooley, Richard and Jackie Fortey, Mary Rossetti Rutterford, Angela Thirlwell, Rosalie Silverstone, Gill Capel, and Bea Fausold. For proofreading and invaluable suggestions: Mark Dalrymple, Ri Renaud, Paola Bonzanini, and Hans-Jörg and Cheryl Begrich. At the U.S. Geological Survey: Peter Modreski, Thomas Dolley, and Lisa Corathers. And my kids, Kate and Iain, who constantly applied the important pressure of asking how the book was going.

This book would, quite literally, have been impossible without Lavinia Trevor, my tenacious agent, and Jenny Wapner and the team at University of California Press, all of whom encouraged and guided me into the world of publishing. My heartfelt thanks to Dore Brown and Jennifer Knox White, who thoughtfully and thoroughly, but always gently, edited the manuscript into a vastly improved shape. Dore wrote that they "tried not to still the occasional meanderings, but simply to free up the larger eddies that threatened to impede forward motion." They succeeded: any remaining turbulence is entirely my own.

I am under instructions to say little about the role of my wife, Carol, but little is precisely what this book would have been without her. Researcher, contributor, proofreader, and interrogator of the *Chicago Manual of Style,* she took on all the tasks that allowed me to *write* the book, including, with not quite perfect consistency, having a Job-like level of patience.

PREFACE

Sand is overrated—it's just tiny little rocks.
Eternal Sunshine of the Spotless Mind

I should have turned left some distance back, but following twisting and diverging tracks in the sand took concentration and made distinguishing a left turn difficult. There were no signposts; the map was a tentative set of branching spidery lines, petering out into not very much. Having missed the left turn and driven straight on for some time, I found myself entering the Rub' al-Khali, the aptly named Empty Quarter, the largest sand sea on Earth; the sight, the sense of the place, was unlike anything I had encountered before—immense, impersonal, and, yes, very empty. I was on my own in an ancient but valiant Land Rover, attempting to catch up with my companions. It's tempting to say I was lost and terrified, but in truth I was neither: if I looked behind me, I could still see, over the tops of the rolling sand hills, the Oman Mountains, out of which I had recently driven. But if I looked ahead of me, there was nothing, nothing but a horizon lost in haze, and I truly did not know where I was—I was not terrified, but I was deeply nervous.

Over thirty years ago, I was on my first job as a professional geologist, surveying the Oman Mountains as part of a British government aid project. The country was being dragged out of the dark ages by its new sultan, who had decided that his father's medieval approach—wearing glasses had been, until recently, illegal—required change. Maps, communications, transport were, to say the least, unsophisticated, and the British Army was still attempting to control insurgency. As geologists, we needed to spend our time in places the army advised us against visiting, and we tended to avoid the scattered villages that marked the occasional emergence of water. We did so, however, not through safety concerns, but simply because we could not afford the time required by the embracing hospitality of the local people, the protocol of moving from one elder's palm-shaded house to another's for yet more varieties of dates.

But now I was a long way from any hospitality, and felt it. Why I was on my own I forget exactly, but it would have been a need to stock up on tinned supplies or flour, or to collect yet another spare part for one of the Land Rovers. I had elected to do this while the others went ahead and set up camp in our next area of work. We camped in wadis, the great dry gullies and canyons typical of any arid land. The wadis were the highways, for us and for the occasional storms that gathered in the mountains and, in flash floods, poured huge volumes of sand, gravel, cobbles, and boulders out onto the edge of the desert. Unfortunately, because of the scale of the work done by the wadis in burying the flanks of the mountains in a chaos of detritus, it was necessary to drive a long way out into the desert before the gullies were shallow enough to allow travel parallel with the mountains. This, together with my missing the left turn, had put me out on the edge of the Empty Quarter, or, as the antique map hanging on my wall objectively describes it, the "Great Desert between Mecca and Oman."

I had, at the time, no aspirations for desert exploration, no desire to follow in the footsteps (or camel prints or car tracks) of Wilfred Thesiger or T. E. Lawrence, and I gave no thought to the basic rules of desert travel (number one being "don't do it alone"). Panic was beginning. I drove on. I lurched over the top of a dune and slid to a halt, faced with two well-armed military types standing in front of a small hut, there in the middle of nowhere. I had, unknowingly, crossed an international frontier, and these were the border formalities for the United Arab Emirates. This was not the kind of human presence I had hoped to encounter, and the panic rose. We had no common language, other than my feeble attempt at a cheerful greeting in Arabic and the universal gestures of pointed guns. I quickly pulled out a letter of introduction from the Oman government to convince anyone who needed convincing of my peaceable and purely scientific intentions. My hopes rose when this was carefully scrutinized, but fell when I realized that the document was upside down. For whatever reason, however, it seemed to work: swirling motions of rifle barrels suggested that I turn around, and this I hastily did, waving goodbye and disappearing back toward the mountains in a cloud of sand. Some hours later, I rounded a bend in a large wadi to see a cluster of familiar tents. I arrived, to the relief of companions who had expected me somewhat earlier, in time for the evening meal of chapatis and something tinned. The following day, I returned to the mundane task of sieving wadi sand for mineral analysis and banging at the mountains, as geologists do, with my back to the Rub' al-Khali.

Deserts are arguably the most dramatic and evocative images of sand for all of us, whether we've experienced their immensity for ourselves or not. Direct encounters with sand are typically at the beach or, more frustratingly, on the golf course or in the process of trying to set level paving stones. But why a book on this seemingly mundane stuff? Because, as I hope to convince you, sand is anything but "just tiny little rocks"; sand is one of our planet's most ubiquitous and fundamental materials and is both a medium and a tool for nature's gigantic and ever-changing sculptures. Because, as William Blake recognized—"To see a world in a grain of sand"—every sand grain has a story to tell, of the present and the past. Because without it, our world, both on a global scale and on the scale of our everyday lives, would be dramatically different. Sand is all around us in different guises—it has made possible the information age and the computer on which these words are being written, it provides the basis for buildings (and windows), it enables us to have toothpaste and many pharmaceuticals, cosmetics, and foods. It is an unsung hero of our world and our everyday lives, as well as, quite frequently, a nuisance. Sometimes it sings to you. Its stories of diversity, power, and beauty deserve to be told, whether they are of individual grains of sand or of the great deserts of this world and others.

As I gathered thoughts and materials for this book, the potential scope began to reveal itself, and the challenge became how to weave the narrative through all the diverse stories of sand. The task rapidly became deciding what to leave out. But it also became clear that the nature of sand lends itself to using, metaphorically, the anthropologist's approach. An anthropologist might begin an account with individuals and their character, building up into a discussion of family units, tribes, and societies, relationships and dynamics, taking note along the way of imagery, imagination, and myths. Thus this book builds up in scale, beginning with sand in the natural world: individual grains (chapter 1) and their strange collective— tribal—behaviors (chapter 2). Then we follow the great journeys and migrations of sand that are driven by water, wind, and ice, together with the landscapes that are both the stage for these journeys and their outcomes (chapters 4, 5, and 6). Much of the Earth's history is recorded in the testaments of ancient sands, and using modern sand as our interpreter, we review some highlights of these records (chapter 7). Then, moving from the natural to the man-made world, we look at the astonishing diversity of ways in which sand is our servant, a critical ingredient of so many aspects of our lives (chapter 9). Along the way, we take a couple of breaks to reflect on how the imagery of sand has embedded itself in our imaginations—in mathematics, art, literature, and language, as medium and muse (chapters 3 and 8). Finally, we peer outward and onward, beyond the Earth and beyond the present (chapter 10).

This book is very much about *scale,* from the microscopic to the global and the astronomical, and from hours to eons. Sand is a character that has a role to play at every scale, both in the physical sense and in our collective imagination. There are worlds to see in a grain of sand, and countless grains to see in our world.

"To measure is to know." Because this book is very much about scale, measurements play an important role. Where possible, I have tried to use evocative and analogous measures, but explicit quantities are unavoidable. Scientific investigation is conducted and reported globally in metric, or International System, units, and the primary measures in this book are thus expressed. Each is followed by a translation in parentheses into United States customary units (otherwise known, quaintly, as *English* or *imperial* units), whose international definitions are, nevertheless, metric.

Individuals

Birth and Character

Who could ever calculate the path of a molecule?
How do we know that the creations of worlds are not
determined by falling grains of sand?

Victor Hugo, *Les Misérables*

For Nature is the noblest engineer, yet uses a
grinding economy, working up all that is wasted
to-day into to-morrow's creation; not a superfluous
grain of sand for all the ostentation she
makes of expense and public works.

Ralph Waldo Emerson, "The Young American"

CONCEPTION

It was love at first sight. As the raven circled over the endless ocean, he saw a beautiful mermaid and was entranced. He flew close and asked her to marry him, and she agreed—but on one condition: "Make me some land where I can sit on a beach and dry my hair and I will marry you." The raven knew he would need help with this task and, in return for not insignificant favors but without revealing his intentions, enlisted the seal and the frog to procure some sand from the bottom of the sea. The raven then flew up into the strong winds above the ocean and scattered the grains to every corner of the world. At the place where each grain fell into the ocean, an island was formed: small islands from the tiny grains, large ones from the biggest grains. The mermaid was delighted, for the first time in her life drying her hair on a sandy beach. And then she and the raven were married.

This creation story, as told by the tribes of the Pacific Northwest whose ancestors were the raven and the mermaid, is only one example of myths from around the world in which grains of sand are the fundamental components of the Earth's creation. From the east coast of North America to the Carpathians, from West Africa to the Pacific Islands, sand is the parent for the birth of the land, a powerful symbol of origins. In *The Neverending Story,* the fantasy novel by Michael Ende, translated from the original German in 1983 and adapted into several films, the land of Fantasia is destroyed, with only a single grain of sand remaining. But that single grain enables the rebirth of the world.

The symbolism of sand as a foundation for our world continues in our collective subconscious, and the story of sand is indeed never-ending.

As we all know from the aftermath of a family visit to the beach, a single grain of sand can get anywhere. It physically penetrates, often to the detriment of health or a piece of machinery, and it can embed itself in our imagination. Pick up a single grain from the beach, look at it through a magnifying glass, and you have embarked on a journey taken by poets, artists, and philosophers—not to mention geologists. William Blake's "To see a world in a grain of sand / And a heaven in a wild flower," from his "Auguries of Innocence," has been put to use countless times to refer to flights of the imagination (the more gloomy direction that the poem subsequently takes is often ignored). Echoing Blake, but in perhaps a more approachable vein, Robert W. Service wrote, in "A Grain of Sand":

> For look! Within my hollow hand,
> While round the earth careens,
> I hold a single grain of sand
> And wonder what it means.
> Ah! If I had the eyes to see,
> And brain to understand,
> I think Life's mystery might be
> Solved in this grain of sand.

What is it about the idea that within its minuteness a grain of sand encapsulates greater things, that it is a metaphor for a grander scale, that it has a story to tell? There is a temptation to anthropomorphize, to gaze into the weather-beaten face of a sand grain and see ourselves reflected, our own life stories, our own journeys, our own worlds, to see the grain as an individual with a *character,* as well as a member of a family and larger clans, extended global tribes. Anthropomorphizing, yes, but it does provide a deep resonance and a framework within which to scrutinize a grain of sand. The birth of a sand grain is a microcosmic event, a flap of a butterfly's wings heralding greater change and a larger creation. Each grain carries the equivalent of the DNA of its parents and develops a character through its life that is molded partly by its parentage, partly by its environment. Compared to the scale of a human life, however, the sand grain's story is never-ending, and rebirth is a regular event.

BIRTH

In order to read the stories hidden in a grain of sand, we need to look at its exterior and interior, to take it apart. Like people, each sand grain is unique but belongs to a particular family with common genetics and origins. Just as stories are told in different languages and emerge from different cultures, sand can be created in different ways and can be composed of a wide variety of substances, although there is one dominant group in the population that shares a common history and a common chemistry.

Sand can be made by simply grinding up rocks into smaller and smaller pieces, but this is not easy and only glaciers do it effectively. Sand can be made biologically, from small shells and other products of the living world; whole beaches are formed this way. Warm seas can deposit their dissolved minerals, like limescale in a kettle, making minute pellets of sand. Sand grains can originate cataclysmically, as when molten rock spewed from a volcano chills and shatters in the air, or as the surface of the Earth melts under the impact of a meteorite, ejecting cascades of liquid droplets into the atmosphere; these solidify and shower back across oceans and land to be found as individual grains within sand or sandstone. But by far the majority of sand grains are made of one of the Earth's most common ingredients, the mineral quartz, and are formed by the process that works, day in, day out, on every exposed piece of land on the Earth's surface—weathering.

The most common element in the crust of the Earth and in the land around us is oxygen. Not as the gas that we need to survive, but chemically bound up with other elements to form solid—and not-so-solid—minerals, just as sodium and chlorine join forces to make salt. The second most common element is silicon, which teams up with oxygen and other common elements, such as aluminum, iron, magnesium, potassium, and sodium, to make the dazzling variety of minerals that are the ingredients of most of the Earth's crust—the silicates. One family of these minerals, the feldspars, is the most common constituent of the crust. But silicon and oxygen themselves make a fine couple, strong and enduring: together, they form the mineral quartz, the common form of silica. Close to 70 percent of all the sand grains on the Earth are made of quartz—tiny crystal balls, each with its own revelations.

Born in the cauldron of the molten depths of the Earth's crust, *igneous* rocks, cooled and solidified in a glittering matrix of crystals, are out of their element when they are ultimately jacked up by tectonic forces and exposed at the Earth's surface. However hard and durable they may seem, rocks such as granite are vulnerable to the weather, many of their constituents chemically unstable. The ravages of time and the elements are obvious on old gravestones and buildings—the corrosion of Cleopatra's granite needles is the classic example. In the midst of that corrosion, sand grains begin to sense freedom.

Chemistry and acidic rain are prime actors in the drama of weathering, but they are not the only members of the cast. Temperature changes, expansion and contraction, freezing, thawing, and the chemistry of water all work away at cracks, even in arid climates. And plants have leading roles, too. Not only does the merciless growth of roots physically tear open the fractures, as in the old tree-lined sidewalks of so many towns, but the roots are chemically active. They do, after all, feed the tree. In conspiracy with minute fungi, the roots extract ingredients essential to the plant's growth—removing them from the minerals among which the roots have worked their way, and removal weakens the rock. It doesn't take mighty trees to accomplish this—humble lichens and algae effectively rot, slowly but surely, the rock on which they live.

All rocks, even the tough ones, like granite and its relatives, rot. The weakest links in the chain are the first to go, and in a granite these are the feldspar crystals and their fellow silicates. Quartz is made of sterner stuff, thanks to its internal structure. In a quartz crystal, the average composition is one silicon atom for every two oxygen atoms, but there is no such thing as an SiO_2 molecule: the silicon and oxygen conspire together to construct incredibly strong chains of pyramids, and the chains interlink—like DNA—in long helix-shaped spirals. This structure is almost impregnable—quartz is a survivor, hard, resistant, and extremely difficult to dissolve. In granite, each crystal of quartz is surrounded by weaker neighbors; other minerals, originally formed under more extreme conditions, are more vulnerable and unstable at the Earth's surface: they corrode rapidly. Feldspars rot away to form clay (the granites of Dartmoor decay to provide the vast deposits of "china clay" historically vital to the ceramics industry). Support for the quartz grain vanishes and, like a loose tooth, it drops out of the rock. A sand grain has been born.

Rotted, corroded, fragmented, pulverized. *Comminuted.* The ultimate fate of the toughest rocks is to be broken into pieces, *clasts* (from the Greek *klastos,* "broken"). The feet of the towering cliffs of Yosemite are draped in the detritus of granite simply falling apart. "The mine which Time has slowly dug beneath familiar objects is sprung in an instant; and what was rock before, becomes but sand and dust" (Charles Dickens, *Martin Chuzzlewit*). The sand grain has become a symbol of impermanence and the fragility of our—and nature's—works.

The birth of a sand grain in this way signifies the death of a mountain. The rocky outcrop from which it fell is now infinitesimally smaller. But the effects accumulate. It has been estimated that on the order of a billion sand grains are born around the world *every second;* add up these seconds over the billions of years of the Earth's history and the scale of change that erosion can cause is clear. But we can also see it happening every day on our time scale. Dramatic changes can happen overnight, as when the Associated Press reported that "New Hampshire awoke Saturday to find its stern granite symbol of independence and stubbornness, the Old Man of the Mountain, had collapsed into indistinguishable rubble" (May 3, 2003). The average effects of these processes over the Earth's surface are difficult to measure, but typically the landscape of a mountain range will be lowered by a few millimeters, a tenth of an inch or so, every year, year in, year out. The processes of weathering and erosion are immensely complex and difficult to measure. The very term *weathering* is probably misleading, since the rate at which it happens does not correlate clearly with weather or climate. It is apparent that much of the corrosive chemistry happens *below* the surface, where rocks are saturated with water moving through fractures large and small, eating away between the mineral grains. But the effects are there for all to see.

Of course, weathering eats away at everything exposed to the elements, not just granite. Sand grains originally born from granite long ago may accumulate, be buried, and become naturally glued together, *lithified* (from *lithos,* Greek for "stone" or "rock"), into a solid *clastic* sedimentary rock, a sandstone. When this, in its turn, is exposed at the surface, it is attacked by weathering and the sand grains are liberated again. The whole process is cyclic, over and over again, each time the grains carrying with them microscopic evidence of their parentage, their genetic origins. The majority of quartz sand grains are derived from the disintegration of older sandstones; perhaps half of all sand grains have been through *six* cycles in the mill, liberated, buried, exposed, and liberated again—as observed by Emerson in the opening of this chapter, reborn repeatedly.

THE IMPORTANCE OF SIZE

Our sand grain, newly born, finds itself, together with a motley collection of other detritus, organic and inorganic, as part of a soil, the *in situ* accumulation from the physical, chemical, and biological processes at work in a particular place. The sand grain is anonymous, waiting for rain and wind to sweep it away on an endless journey, to demonstrate its durability while its weaker companions fall by the wayside. But it is called *sand* not because of what it is made of or its origins, but because of how big it is.

Sand is somewhat like beauty—we know it when we see it, or touch it, but it seems difficult to describe. However, if we are to understand it, to use it, to live with it, then we have to tackle this problem. A U.S. geologist, Chester K. Wentworth, took on the task in the early twentieth century. The first sentence of his publication on describing sand reads: "In no other science does the problem of terminology present so many difficulties as in geology."

What Wentworth set in place was the concept that the only thing that matters is particle size: composition is irrelevant. This has proved an enduring and important approach. The behaviors of anything made up of relatively hard bits and pieces of a particular size, regardless of what the bits and pieces are made of, are unique and, in the case of sand, quite odd. The sugar in your teaspoon, poised over your cup of coffee, is, technically, sand. In the coffee it may not last long, but if you poured it over the kitchen floor and began blowing over it, you could begin to make sugar dunes. The salt in your salt grinder starts off as coarse sand and, when you grind it, becomes fine sand. It may dissolve in the saucepan, as does the sugar in the coffee, but until it does, it's sand. Indeed, there is a desert with massive glistening-white windblown dunes made of something very much like salt: the dunes of the White Sands National Monument, in New Mexico, are made of crystals of the calcium salt gypsum, formed from the evaporation of desert lakes—but the sand grains dissolve when it rains. And along the shores of the lowest point in Africa, Djibouti's Lake Assal, the water is so saline that the sand is made of salt crystals.

So, size is what matters. But how to define size? Sand grains come in a variety of shapes, which can make measuring size quite tricky. Think of tomatoes. How would you measure the size of a tomato, compared to another tomato? Certainly, many are roughly spherical, but they are also commonly rather wrinkled, some with deep and contorted folds; some are more pear-shaped, and some are obviously oval, with some of these being essentially long and thin. What they taste like—what they are made of—doesn't matter in this instance; it's simply a question of describing their *size,* and how is this done among objects with such varied forms?

The idea of measuring size works well only for regular objects, the obvious examples being spheres and cubes. The diameter of a sphere or the length of the side of a cube defines each one. But for an irregular object, its size depends on *how* we measure it. In turn, how we choose to measure it depends on *why* we want this information. Why should we want to describe the size of sand grains? From a scientific point of view, we want to understand, for example, how dunes form and what determines the shape of a dune in a particular place at a particular time. How do dunes move? How do sand grains get picked up and hurled around in a sandstorm or in a flash flood in the mountains? How do water and air compare in their ability to move materials of different sizes? We are interested in why rivers form the meandering shapes they do, and how the sandbars of ancient rivers from hundreds of millions of years ago are preserved and what they can tell us about conditions on our planet back then: all good geological science. But measuring the size of sand grains can also be important on a day-to-day, more "practical" level. The size of the sand in concrete or tarmac makes a difference to its strength and other properties; when sand is used as a filter, the size, and the range of size, of the grains is fundamental. Much of our water supply comes from underground sands whose grain size determines how much we can extract and how much will be replenished. How effectively we can maintain harbors, dams, rivers, and coasts depends on our understanding of sand movement, and movement depends on size. Measuring the size of a sand grain is critical to a surprising number of aspects of our daily lives.

SCALES

Wentworth established the scale of sand grain sizes that, essentially, is still in use today. But to do so, he refined and documented an earlier scheme proposed by Johan August Udden in 1914. Udden emigrated as a child from Sweden to the United States with his family and as an adult was fascinated by natural history. Like many natural historians of that era, he dabbled in different things, but his name lives on in the Udden-Wentworth grain size scale. Udden was obsessed with sand. He collected sand, as well as dust and pebbles, from a wide variety of locations, and measured the size of the grains. The data were published in an exhaustively long list that reveals the eclectic nature of his collecting. Samples of "sand blown on a snowdrift, Baltimore, Maryland" were carefully measured. Dust samples "collected in a running railroad coach in North Dakota after a storm," "from the top cloth on a flagpole, August 19, 1895, Rock Island, Illinois," and "washed from the leaves" of hickory, linden, and oak trees were all documented.

Most of Udden's measurements were made using the simplest tools: a series of sieves. The sand is put through the coarsest and then successively smaller and smaller mesh sizes, each fraction being weighed. The minimum size of the hole that allows a grain to pass through is then compared with the hole size that stops it, and the grain's size defined as the average of the two. It's a tedious and backbreaking business if automation is not available.

The Udden-Wentworth scale defines the terminology and size categories for different grain sizes, from microscopic particles of mud to boulders (the latter being measured directly, not sieved). It clearly averages out sizes for grains that are more or less oval—the shape of the grain has to be considered separately. Wentworth recognized that using a linear scale, like that on a ruler, would be impractical, given the huge range of sizes involved, and that it would distort the signifi-

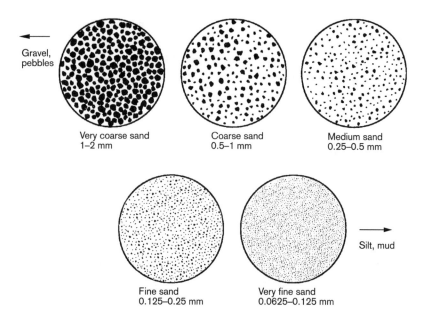

FIGURE 1. The classification of sand by size, after Chester K. Wentworth's original illustration, 1922. The scale shown here is accurate. The finest sand grains ("very fine") are .06 millimeters ($\frac{1}{400}$ in) in diameter; the largest ("very coarse") are 2 millimeters ($\frac{1}{12}$ in).

cance of size *increments:* an increase in size equivalent to the thickness of a human hair makes little difference to a pebble but could double the size of a sand grain. The sensible and practical way to address this would be to use a *geometrical* scale, one where each division boundary is a multiple of the previous one. As Wentworth wrote: "The use of a geometrical scale makes the successive grades fall into equal units on the graph—an arrangement much easier to read and interpret than any other known to the writer." This was, in fact, an early recognition of an important way in which nature works—on vast ranges of scales that are fundamentally geometrical, not linear, based on ratios rather than size for the sake of size.

The basic ratio in the sand size scale is two: each division is twice the previous and half the following as sizes increase. So, after all this, what does the scale of grain sizes look like?

Figure 1 is redrawn from Wentworth's original publication and the scale is accurate: the grains appear as they would scattered on the sheet of paper. The smallest grain that can be called sand is essentially invisible to the naked eye as an individual grain; a magnifying glass becomes more and more necessary the further below "medium" one goes. The largest sand grain is 2 millimeters ($\frac{1}{12}$ in) across, but the majority of the world's population of sand grains measure around 1 millimeter ($\frac{1}{24}$ in).

A single layer of grains of very coarse sand covering an area the size of your fingernail would contain around thirty grains. A layer of very fine sand would contain twenty-five thousand grains—sand covers quite a spectrum of sizes. This range reflects the genetic inheritance of the grains—each is the size of the crystal that naturally constituted its parental rock. But the behavior of the sand family of grain sizes is very distinctive and different from that of its larger and smaller relatives, gravel and silt. Sand grains, when gathered en masse, form a very strange material, as we shall see in the next chapter.

FALLING SAND

There are other, more esoteric ways of describing grain size. The principle that the definition used depends on the purpose is illustrated by that particular group of scientists and engineers who are interested in how sand grains move, through air or water. Drop a spoonful of sand into a jug of water, and the grains will drift downward at a constant speed: the pull of gravity is balanced by the resistance put up by the water. If your sand contains grains of different sizes, the bigger, heavier ones (assuming they are all made of the same material) will settle faster since the resistance of the water affects them less. The result will be a layer of grains on the bottom of the jug that contains the coarsest ones at the base and the finest at the top. This effect can be seen in ancient sands now thoroughly solidified into rock. A sweeping current will flush along a range of grain sizes, and when it stops, the biggest grains settle first. The sand is graded; nature has carefully sorted out the sizes and dumped them in sequence on top of one another (Figure 2, left). This illustrates another of the important characters of families of individual sand grains: the *range* in size within that family. Pick up a handful of sand at the beach and look

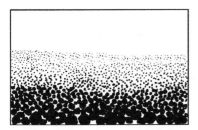  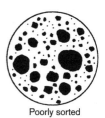

Well sorted   Poorly sorted

FIGURE 2. Graded sand (left); examples of well sorted and poorly sorted sand grains (right).

at it closely. Even to the naked eye, it's clear that while most grains may be roughly the same size, there are also smaller ones and bigger ones. Udden referred to the "law of the chief ingredient": in most sediments, the average size of grain is present in greater quantity than any other size. But the *range* of grain sizes is an important clue to the story of a sand family. By carefully sieving a sand through successively smaller sieve sizes, the proportion of grains of each size range can be measured. Families of grains that are all roughly the same size are called well sorted; those that comprise a wide range of sizes are poorly sorted (Figure 2, right).

If you were to drop your sand through the air, the grains would obviously fall faster than they did in water—the resistance of air is much less. But the ability of fluids, which, in our world, are generally air, water, or ice, to pick up and carry off sand grains of different sizes is key to fundamental geological processes. As will be discussed in later chapters, sand grains as individuals, families, and tribes love to move and to congregate—as sediment. The way in which they move is determined by the size of the grains, their shape, and the medium through which they are moving. It's therefore logical that those people interested in these processes of sand movement, on all scales, should define grain size in terms of how fast a grain falls through a fluid.

For some researchers, like Ralph Bagnold, their whole definition of sand is based on this. An extraordinary man, Bagnold had two remarkable careers, one in the military, the other in science: both brigadier and fellow of the Royal Society, he was a professional soldier and an "amateur" scientist driven by curiosity and armed with enthusiasm and a huge intellectual capacity for both practical and theoretical analysis. In his careful and quantitative approach to how sand moves, Bagnold made groundbreaking contributions from the 1930s to the 1980s. In the desert, he defined the lower limit of what could be called sand as grains that are too big to be picked up "by the average surface wind," suspended, and swept away as dust. In other words, the speed at which the grain falls is too slow for it to compete with the wind and settle on the ground; it stays in the air. The upper limit is anything that is too big to be even nudged along the ground by the wind or the impact of flying grains. The implication of this is that windblown sand in the desert should only be of a limited range of size, purely because of the physics of moving it around, and Bagnold showed, through his own and Udden's measurements, that this is indeed the case. Much of the sand in the deserts of the world is "very fine" (Figure 1). Bagnold recorded the minimum wind speed needed to start sand grains moving, a critical factor in explaining how deserts operate (more on this in chapter 6).

WATER

Obviously, water is a different matter from air—it's much more dense and viscous and is more effective than air at moving bigger grains (for example, boulders) around. But the principle remains the same. The movement of sand by water is an incredibly complicated process; Albert Einstein is reputed to have warned his son against becoming a river engineer because the physics of sediment transport is too complex. As with many natural processes, we can analyze some aspects and make complicated real and mathematical models of simplified situations, but we still do not understand it completely. Having tired of wind-blown sand (only because of the absence of accurate long-term wind data), Bagnold turned his attention to water and set the science firmly on its path. But it still has a long way to go. We will return to this thorny topic, but for now, a taste of the strange behavior of sand and why size is important:

Imagine that you have put on your scuba gear and have anchored yourself to the bed of a river. Beneath you is the riverbed, made of nothing but mud. The current of the river flows clearly over the bed, disturbing nothing. Suddenly, you feel the current speeding up—a storm upstream in the hills has poured volumes of water into the river and its flow has increased. But the mud remains undisturbed. Buffeted by the flow, you are suddenly surprised to see myriad grains of sand, a lot of them quite coarse, sliding, skipping, rolling, and jumping along the bed beneath you. You look more closely, and bouncing grains hit your face mask, but the mud remains undisturbed. Further upstream, the increased current has torn the side out of a sandbank on a bend in the river and is moving the debris downstream. But the mud remains undisturbed. Why? This seems to fly in the face of logic: larger material being mobilized while finer material stays where it is. You decide to retreat to the riverbank before the cobbles and boulders arrive.

Secure beside the surging river, you think this problem through. There are some simple rules in nature, but often there is a point at which simple rules stop working. It would seem obvious that if it takes a certain speed of current to move a particular size of sand grain, that current will also move anything and everything smaller. But your subaqueous observations conflict with this. The microscopic particles of mud and clay glue themselves together and resist being picked up by the current, which instead simply flows over them. Of course, if you had dug your hand into the mud and disturbed it, it would have been flushed away in an instant, but once mud has settled and consolidated itself, it's extremely difficult to budge— so difficult that doing so can take water flowing at the same speed that is required to shift a boulder a meter (3 ft) across. But in between, easily shifted at far slower speeds, is sand. Of the entire range of grain sizes in the Udden-Wentworth scale, sand grains are the most easily moved, which explains why there is so much of it in so many different places and why so many landscapes are constantly on the move.

The endless journeys that sand grains take age them. Sharp edges and angular corners are knocked off even durable quartz grains by the constant battering as grains collide with one another. However, once a grain has become smooth and round, it stays that way, with little further change, for enormous lengths of time. Wind is immensely more brutal than water in rounding off grains, the violence of the impacts being hundreds of times more effective than those cushioned by water. The world's deserts are the source of virtually all neatly rounded sand grains; it would take journeys down thousands of Mississippis to achieve the same effect.

Quartz grains may be the ultimate survivors of these journeys, but quartz is far from the only ingredient in the sands of the world. In fact, the dazzling variety of sands makes them collector's items.

ARENOPHILIA

Anything can be collected, and people collect anything. Stamps, beer cans, beetles, postcards, garden gnomes, chocolate bar wrappers—and sand. There are countless sand collectors around the world, and there have been for a long time. Today, they have websites (often with stunning photographs), chat rooms, and a market for the constant exchange of samples. Sand is, after all, a relatively simple thing to collect—it's free (excluding the cost of travel), found almost everywhere, compact and easily storable in small glass jars, and endlessly diverse.

Some sand collectors gather their samples for emotional reasons. It has become a tradition for survivors of Iwo Jima to revisit the island and take some of its sand home. Other collectors enjoy the unique character of sands from different places, the colors, shapes, and textures. An array of sand samples almost looks like a box of colored crayons, muted perhaps, more of a landscape painter's palette, but a dazzling range of hues. Even an apparently nondescript brown sand, if looked at closely, reveals its own glittering, granular character. A simple hand lens or, even better, a cheap microscope will open the door to worlds of extraordinary beauty and variety. Sand collectors call themselves *arenophiles,* or "sand lovers," from a mixture of Latin and Greek. The word *arena* derives from the ancient Roman habit of covering the ground in amphitheaters with sand (*harena* or *arena* in Latin)—to soak up blood. The pure Greek would be *psammophile,* and some sand collectors use this, but it is commonly used also to describe plants and creatures that are sand-loving, forging a livelihood among the grains.

An old friend of mine in California is a professional geologist and an arenophile; when he learned that I was working on this book, he sent me samples of his collection, set out in a pill-organizing container, a sand for each day of the week (Plate 1). These sands are samples from his travels in Florida, Sumatra, Algeria, Mexico, Tahiti, Bali, and the Galapagos. To this palette could be added sparkling green, deep red, true yellow, purple—a spectrum to gladden William Blake's eye. They are all sand, regardless of color, shape, composition, or origin, simply because the grains fall within the size range that defines sand.

The visually stunning spectrum of sand materials and colors is one of the factors that make sand collecting attractive. Displays of sands can seem like works of art, and works of art can exploit the character of sand—see, for example, some of the creations of Nikolaus Lang. Loes Modderman is a Dutch arenophile and an enthusiast for all things microscopic; her website contains images that are beautiful in their own right (Plate 2).

In the creation myth of the Shilluk people of Sudan, the diversity of the Earth's sands explains the diversity of its people: their creator, Juok, wandered the Earth, and as he did so, he shaped white people from white sand, red and brown people from the mud of the Nile, and the Shilluks from the black Earth around the White Nile.

INGREDIENTS

Quartz—otherwise known as silica, SiO_2—is the potato, the staple ingredient, of sand cuisine. Some sands are made of essentially 100 percent quartz—the top three in Plate 1, for example, which were taken from beaches and a desert. Other sands contain no quartz at all. We know that sand is simply a matter of size; therefore *anything* reasonably hard that presents itself as a sand-sized grain is entitled to form a sand. And a remarkable variety of things do.

A traveler expects to sample a local cuisine that has its origins in local ingredients, and it's no different with sand. Relaxing on a beach in North Carolina or the south coast of England, looking at the sand between your toes, you would hardly expect it to be made of bits and pieces thrown out of a volcano. Trekking along the coast of Greenland, you would be surprised if the sand were composed of the debris from a coral reef. What you *do* find are sand grains of predominantly local parentage. On the Normandy beaches where D-day landings took place, you will find sand-sized fragments of steel.

More often than not, each grain of sand is a member of a microcosmic family of local minerals, with local rocks providing the supply. Look closely at an everyday beach sand from the Isle of Wight, off the south coast of England (Plate 3, left): a nondescript brown sand becomes varied and sparkling under the microscope. Apart from clear grains of quartz, you see a variety of other, more colorful grains—different minerals, fragments of rock. On some beaches, the local ingredients can create dazzling collections of jewel-like grains—for example, the garnets from the coast of Provence in France in Plate 3 (right). These tiny grains are not the kind used to make jewelry, but they are certainly good for sandpaper.

Any kind of rock can provide the material for sand. Some of Hawaii's famous beaches are made up entirely of grains of, unsurprisingly, lava. Where the molten flows enter the sea, they are shattered by the thermal shock of hitting the seawater and then further pounded by the waves. The quenching of the molten lava often takes place so quickly that natural black glass is formed. The results are the famous black sand beaches, bizarre and stark places where jagged fragments of lava protrude from somber sands. Look at the sand grains closely, and the remains of the minute gas bubbles that were solidified into the lava can be seen. In places, the lava sand becomes so coated in iron oxide, rusted by the elements, that it takes on a deep red hue. But arenophiles beware: the volcanic goddess, Pele, is reputed to resent the removal of any of her landscape and is said to curse those who do so. Folklore or a taxi driver's tale this may be, but the U.S. mail has delivered envelopes of returned sand to Hawaii, together with apologetic notes.

Quartz may be the most durable of the Earth's common minerals, but many of its brethren are no wimps. A host of different individual minerals form individual sand grains. Many of them are quite valuable—the diamonds of the Skeleton Coast of Namibia, for example. Along other parts of Hawaii's shores, the apple-colored mineral olivine, another product of the volcanoes, survives well enough to construct startling beaches of green sand.

It has long been known that magnetic iron minerals can make up a sufficient proportion of a sand to affect a compass. In 1733, Petrus van Muschenbroek, professor of "Mathematicks and Astronomy" at the University of Utrecht in Holland, wrote to the Royal Society in London of his experiments with "Magnetick-sand," describing samples from Virginia, Persia, and Italy in terms of how "attractive" they were.

One of van Muschenbroek's forebears among Dutch scientists was Antony van Leeuwenhoek, an early and meticulous arenophile. Born in Delft in 1632, van Leeuwenhoek had endless curiosity but no wealth and no higher education. He did, however, have wide interests, learning the art of grinding glass and eventually making his own microscopes, more than five hundred of them. As we shall see in chapter 9, the technology of turning sand into glass underpinned the Renaissance.

The effect of the new high-quality glass, which allowed the examination of the very distant and the very small through telescopes and microscopes, was felt most dramatically in the sciences. Many of the great scientists of the Renaissance—René Descartes, Robert Hooke, Christiaan Huygens, Isaac Newton, and, of course, van Leeuwenhoek—were skilled glass grinders.

Van Leeuwenhoek's microscopes were far simpler devices than others already in existence, but his were capable of a significantly higher magnification. Through his lenses, he could examine clear, bright images, magnifying two hundred times whatever he put beneath them—which was anything and everything. His list of carefully documented discoveries is a long one; in his own words, "whenever I found out anything remarkable, I have thought it my duty to put down my discovery on paper, so that all ingenious people might be informed thereof." He was the first person to describe bacteria, extracted from between his own teeth. He documented algae and other microscopic creatures, bee stings, blood cells, living spermatozoa, minerals, fossils—and sand.

Van Leeuwenhoek was the first scientist to document the wide range of character displayed by individual sand grains. Perhaps originally interested in the sand he used to grind his lenses, he went on to obtain samples of different sands and subject them to his microscopic scrutiny. He wrote copiously over the years to the Royal Society of London, his letters translated into English or Latin and published in the society's *Philosophical Transactions*. In 1703, by then a fellow of the society, he published a letter "concerning the figures of sand." In this delightful illustrated piece (Figure 3), he begins: "I remember I have formerly affirmed of Sand, that you cannot find in any quantity whatsoever two Particles thereof, that are entirely like each other, and tho perhaps in their first Configuration they might be alike, yet at present they are exceedingly different." He described how sand grains might be abraded and rounded over time and how he had tested their durability with alchemical treatments of fire and acid. But he also had an overactive imagination, creating miniature scenes from the features within individual grains (Figure 3, right): "you might see not only, as it were, a ruined Temple, but in the corner of it GHI appear two images of humane shape, kneeling and extending their arms to an Altar that seems to stand at a little distance from them."

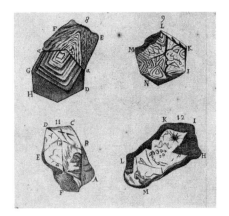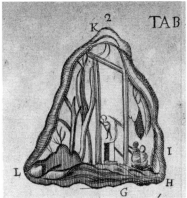

FIGURE 3. Drawings of sand grains from Antony van Leeuwenhoek, 1703. (Image © The Royal Society)

Among the myriad microscopic creatures first described by van Leeuwenhoek were members of the teeming population of the oceans—which themselves contribute spectacularly to the world's sands.

CREATURES GREAT AND SMALL

The living world contributes huge amounts of *biogenic* sand to coasts, beaches, barrier islands, and shoals. Particularly in regions rich in marine life at every scale, from coral reefs to minute floating organisms, much of the sand is the debris of this activity. Many of the world's prized tropical beaches, which are often far from major landmasses, are made up almost entirely of broken pieces of shell, coral, and the other hard parts of marine creatures. The inhabitants are the only source of the sand.

The cast of characters is huge, but among them are some extraordinary and beautiful creatures. The *foraminifera,* or, more colloquially, forams, are a widespread and remarkably diverse group of single-celled organisms. Over four thousand species are currently found all around the world, from the poles to the tropics and from the deep oceans to shallow lagoons; some are found in freshwater or salt lakes. The artistic diversity of the shells they build makes it difficult to believe that they are

FIGURE 4. Sand grains made of foraminifera shells. (Photos by author)

all, in fact, related, as does their range in size, from fine grains to large coins. But however variable and diverse they may be, they are clever and well adapted—they have been around for at least 550 million years.

Forams that inhabit the oceans are very particular about where they live. Some species float, some live on the seabed, and all are fussy about temperature. Their shells can be simple or complex, all of them of exquisite and delicate design (Figure 4). These shells survive after the foram dies; they sink to the seabed or are carried along by waves and currents as sand. Some beaches are made up almost entirely of foram shells. On the left in Figure 4 are foram shells from a beach in Bali (see also Plate 1) that, akin to ball bearings, make walking there in bare feet quite painful. One of the most exquisite and hauntingly beautiful forms of foram sands are the "star sands" of islands off the south coast of Japan, shown on the right in Figure 4. An old Japanese folktale describes how the Polar Star and the Southern Cross decided to bring life to Earth and were sent to one of the islands, where the sea was calm and warm, to give birth. Southern Cross duly produced thousands of children, but the Seven Dragon god of the sea was angry at not having been asked permission. He swallowed the babies and spat out the dead bodies, which floated ashore and formed the beaches of the island. A kindly local goddess gathered them up and put them in her incense burner so that they could join the smoke and be reunited with their mother in the sky. This is why there are so many stars around the Southern Cross and why local villagers still put sand in their incense burners.

In a strange twist, some forams actually *make* their shells out of sand grains, gluing them around themselves to provide protection. They are remarkably particular, selecting only certain kinds and sizes of grains.

Another strange group of grains—those producing gigantic banks and shoals in the tropics and elsewhere—are the ooliths (Plate 4). The two "o"s are pronounced separately, like "uh-oh" backward. The word means "egg stone," from the Greek, and that's what they look like, tiny white eggs. Wherever the water is warm, shallow, and charged with dissolved minerals, currents will gently roll small grains around. As they roll, minerals will precipitate on the surface of the grains, coating them and increasing their size until they are too big to roll anymore. The results are sand-sized grains (almost by definition, since they owe their origins to gently moving water), most commonly made of calcium carbonate, or limestone. The enormous Grand Bahama Banks have been built up over time by the accumulation of shoals and banks of ooliths and other carbonate material of biological origin. From the air, the banks seem to be swept by a delicate floating fabric of silk, festoons of ooliths.

On an island off the coast of Turkey, there is a stunning stretch of white sand known as Cleopatra's Beach. The story is that Mark Antony shipped barge loads of sand to the island to create this stretch of beach for his lover. There are no other beaches like this on the island, made, as it is, of exotic creamy white ooliths. Modern analysis has shown that the sand of Cleopatra's Beach is identical to that forming beaches west of Alexandria on the Egyptian coast; it probably took Mark Antony around sixteen Roman barges to deliver his exotic gift.

Before we move on from this brief sampling of the astonishing spectrum of materials that can make up sand, let us stop for a moment to consider the humble, but often gaudy, parrot fish. The parrot fish gains sustenance by chewing on coral reefs, ingesting the algae and small pieces of coral, and defecating the leftovers as pellets. Large fish can generate over a ton of sand-sized pellets every year. There are many beaches, for example on Hawaii, where, you might be interested to know, you are walking through piles of, well, fish excrement.

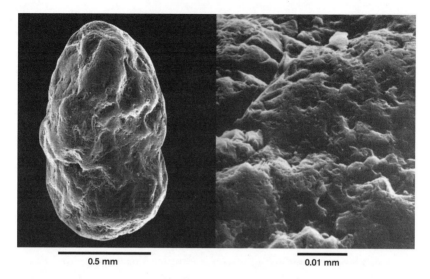

0.5 mm   0.01 mm

FIGURE 5. The face of a sand grain, 0.5 millimeters (¹⁄₅₀ in) across (left); part of the same grain, magnified 1,500 times its original size (right). (Photos courtesy of David Krinsley)

UP CLOSE AND PERSONAL: FACIAL EXPRESSIONS

Close examination reveals many of the details—and the beauty—of sand. This can be taken several steps further with the power of modern technology. Enormously powerful scanning electron microscopes allow the individual, personal features of each grain to be scrutinized, its facial features interrogated for what stories they can tell us about its history. This can, of course, rapidly become a very esoteric subject, but it is one to which many geologists devote their lives.

The face of a typical sand grain is weather-beaten. Pitted and wrinkled, its features show the ravages of time—lots of it. Most sand grains have been through the mill, the geological grinding and battering by wind, water, and ice over eons of time and endless itineraries.

The tiny sand grain on the left in Figure 5 has a distinct physiognomy. It has an oval, rounded profile, with no sharp edges or corners; it has been battered and scarred by countless collisions with its fellow travelers. Look at it even closer (on the right in the figure), and it begins to take on the appearance of an alien landscape, with valleys and ridges sculpted by violent forces; here, the magnification is equivalent to enlarging your thumbnail to the size of a tennis court.

The physiognomy of every sand grain tells a story of its origins and travels. As sand is carried along by rivers, waves, wind, ice, each process imprints a different record, a texture, on the surface of the grains. The ways in which a grain is battered and broken by impacts during a sandstorm are different from those caused by a flooding river. During each period of rest, whether in a dune, a riverbank, or the soil of a garden, different processes operate on the grain and leave their record, sometimes overprinting, sometimes deleting previous textures. Grains can be naturally painted, coated with a thin mineral layer—the black sands of Hawaii turned rust red, for example. A grain may have spent a large part of its life entombed in a sandstone deep below the surface of the Earth, squeezed, heated, and gently, or not so gently, cooked. As the grain's surface becomes chemically etched, perhaps partly dissolved, the hallmarks of this phase of its life often remain recorded even after its reliberation into the hectic world of rivers and waves. Sand-grain detectives can reconstruct stories of each grain's life history.

Every sand grain in the world is unique, an individual. Looking closely at any one of them, diagnosing its condition, reconstructing its life story, and determining its origins is not only scientifically interesting, but also useful. Families and tribes of sand grains from the same place tend to show the same features, a kind of common genetic history. This enables another kind of detective work, the *forensics* of sand, tracking down places and environments of origin.

In late 1944, balloons 9 meters (30 ft) in diameter appeared in the skies above the United States. Landing from the West Coast to Michigan, they carried a deadly cargo: incendiary bombs. Although the only casualties over the following months were, tragically, members of a Sunday school group attempting to retrieve one that had landed, the potential danger to life, towns, and forests was considerable. It was apparent that the weapons had blown in from the Pacific, but where had they been launched? The devices had an automatic altitude-regulation system, releasing hydrogen or ballast to maintain height. The ballast bags were filled with sand. The U.S. Geological Survey's Military Geology Unit, established in 1942, was tasked with identifying the sand. The family of grains was consistent from one retrieved ballast sample to the next, and unique. Distinctive forams and other microscopic shells, together with small amounts of unusual mineral grains in among the granite debris, correlated precisely with beach sands described in prewar geological reports from two locations on the east coast of Japan. Air photographs identified hydrogen production plants at these locations, which were then targeted and destroyed.

The forensics of sand has proved widely useful in archaeology, establishing not only locations and environmental conditions of ancient sites, but also the origins of materials used in tools, carvings, pottery, and paintings. Sand forensics has also been successfully applied to modern criminal detective work.

THE SCENE OF THE CRIME

With the sophisticated microscopic diagnostics now possible, the character of soil and sand as evidence in a wide variety of criminal cases has taken on increasing significance. There are crimes that rarely make the headlines, such as cactus smuggling, that can be routinely solved by pointing to the origin of sand clinging to the roots of the contraband. Investment scams where evidence for a new gold prospect is "salted" with grains of gold from elsewhere can be uncovered by a microscopic look at those grains.

A significant amount of the world's gold supplies comes from the sands of ancient and modern rivers. In 1997 a shipment of these grains of gold worth $3 million was made from mines in the interior of Ghana to the coast and then on to London for processing. After a dispute over the arrangements and cost, the shipment was moved on to Canada via Amsterdam. Canada was the first place where the crates were tagged and given new seals. When they were eventually opened, they contained ordinary sand and iron bars. Where on the shipment's circuitous route had the substitution taken place? The sand was examined by Richard Munroe, a Canadian forensic geologist and policeman. If the substitution had been made in London or Amsterdam, the sand would likely bear the imprint of its northern European origins—particularly the action of ice from the glaciers that had so recently sculpted the continent. But none of those signs were there. Instead, the grains bore the distinctive features of being subjected to a tropical climate, and their composition was typical of the geology of the interior of Ghana. While local security difficulties prohibited making an exact match of the sand grains, any Canadian involvement was ruled out and the insurance claim filed by the mining company was dropped. Sand is a popular material in crimes of "substitution"; in the lively commerce between North and South America, sand has been substituted for, among other goods, cigarettes going south and perfume going north. The genetic fingerprint of the sand involved has pinpointed the location of the crime and helped prove innocence and guilt.

Sand and soil found in the soles of shoes, on clothing, or on tires can place people or vehicles in a particular place—however much they may deny it. Geology has become a standard tool in the kit of government forensic laboratories the world over, but it has been around for some time. The fictional Sherlock Holmes claimed to be able to describe an itinerary from mud splashes on trousers. In real life, evidence from sand has been used for over a hundred years. In 1908, in Bavaria, a poacher was suspected of murdering a young woman. His wife had cleaned his shoes the day before the murder, but they now had three layers of sand and soil stuck on their soles. As part of the investigation, one Georg Popp, a local chemist, applied his geological expertise to these layers. He reasoned that the layer next to the sole of the shoe was the oldest; it was made of the same materials as those outside the suspect's house. The second layer contained red sand and other materials identical to those from where the body had been found. The last and most recent layer contained brick fragments, cement, and coal dust that matched samples from where the suspect's gun had been found. What this layer did not match was the soil from the fields where the suspect claimed to have been walking at the time of the murder. The prosecution case was complete.

On a dark, rainy night in September 2002, a black truck parked beside the Shenandoah River in Virginia. Another truck pulled up, and the window rolled down to reveal the barrel of a shotgun. The driver of the first truck was killed at point-blank range. The murderer left in a hurry, the wheels of his truck spinning in the sand and gravel. After a preliminary investigation, the police had a suspect but insufficient evidence to prove guilt. When the suspect was seen starting to wash his red pickup truck, the police swooped. The truck was spattered with fresh mud: time to bring in the forensic geologists. The mud contained some very distinctive sand grains, a variety of minerals that could only have come from a local quarry. While the quarry was not where the murder had taken place, water washed debris from the quarry into the river, which carried it downstream, mixing and diluting it with the other sand and mud in the river. At low water levels, these were dumped in sandbanks along the river's edge. Geological sleuthing demonstrated that each successive sandbar downstream from the quarry contained less and less quarry debris, and the only one that precisely matched the material from the suspect's pickup was the scene of the murder. The suspect pled guilty in the face of this incontrovertible evidence.

Forensic geology has played a part in a wide range of criminal cases worldwide, but perhaps the most high-profile, yet disappointing, example was the murder of the Italian prime minister Aldo Moro. In May 1978, the body of the kidnapped prime minister was found in a car in Rome. Sand from his clothes and shoes, and from the car, was shown to have come from a particular stretch of beach near the city, yet searches of the area provided no evidence. Other forensic work confirmed the association with this beach, yet the connection with the suspects could not be proved. Years later, the kidnappers declared that they had planted the beach sand as a decoy—whether this is true or not remains unclear.

The world's first database of sand grains has been assembled from soils in the United Kingdom, specifically for police forensics. This database contributed key evidence for one of the country's particularly appalling recent criminal cases, the murder of two young Cambridgeshire schoolgirls in 2002. Once again, distinctive soil under the murderer's car tied him to the location where the victims had been buried.

DEEP TIME FORENSICS

Rain had fallen on the sand, and the small person left clear footprints. Later, the wind rose and filled the prints with sand, covering and burying them. So they remained for 200,000 years until the covering layers were worn away and the prints were discovered by construction workers in South Africa. They are the oldest known footprints made by *Homo sapiens*. Originally thought to be much more recent, they were dated by analysis of individual sand grains through a remarkable technique known as *luminescence,* or *exposure dating.*

Cosmic rays constantly bombard the Earth, at a roughly constant rate. Individual atoms in material close to the Earth's surface are targets for this radiation, which, in the case of a silica sand grain, converts the nuclei of some silicon and oxygen atoms to new, unstable forms, or isotopes, by causing the ejection of electrons that are then trapped in the crystal structure. The longer the grain stays close to the Earth's surface, in soil, sand, or rock, the more isotopes and electrons it accumulates. Measuring the amount of these products in individual grains by stimulating them to release the electrons through luminescence tells us how long the grains have been exposed, or how long it's been since they were buried and shielded from cosmic radiation. There are several variations to this technique, but it has opened up the ability to discover the the age of materials for which no means of dating was available before. Essentially any sand grain can be dated this way. It takes microscopically detailed measurements of thousands of grains to come up with a date, but this method has revolutionized archaeology and our understanding of the history of the Earth's surface.

Prehistoric cave paintings are extraordinarily difficult to date, but occasionally, and with the help of luminescence dating, the archaeologist gets lucky. The Kimberley region of northwestern Australia is rich in art, often painted using ochre pigments that contain no datable materials. But wasps built nests over a couple of paintings in that region, using sand grains as part of the nest structure. The nests were preserved by minerals precipitated from water running through them, and the sand grains could be extracted and dated. The wasps were busy seventeen thousand years ago; the paintings themselves, therefore, must be older than that, making them potentially the oldest pictures of the human figure in the world.

Determining the age of ancient pottery, tools, and carvings is among the many revolutionary uses of this technique. Applied geologically, it tells us the ages of meteorite impacts, fault movements, cave systems—and the landscape itself. Because we can measure how long a sand grain has been exposed at the Earth's surface, we can understand rates of erosion and the durability of landforms. The Atacama Desert in Chile is one of the driest places on Earth and has been that way for a long time. What was not known, until luminescence dating techniques were developed, was quite *how* long the desert's aridity has lasted. Areas of the surface of the Atacama have been virtually unchanged for over twenty *million* years, far longer than previously thought. The desert has witnessed the building of the Andes.

Luminescence dating cannot take us back into the farthest reaches of "deep time," close to the Earth's beginnings, but sand grains do reveal clues from that far back. We know from the ages of meteorites left over from the birth of the solar system that the Earth was formed close to 4.6 billion years ago. For a long time, it was a fiery coagulating mass of molten material, constantly bombarded by incoming debris. This first eon in the planet's history is referred to evocatively as the Hadean, a hellish time. Any material that solidified early on was recycled; the oldest rocks we have found are from northwestern Canada and are "only" a little over four billion years old. But large areas of terrain almost that old are found in Greenland, Michigan, Swaziland, and Australia, and within these rocks are imprints of older events. Water was critical to the early Earth, helping form the atmosphere, rain, rivers, oceans—and life. It had long been assumed that water in any significant quantities, condensing from the steam erupted from volcanoes and arriving extraterrestrially via comets, had taken a long time to accumulate; it was thought a "hospitable" climate developed only after around 700 million years of turmoil. But some very special sand grains now tell a different story.

Western Australia has been the stage on which an immense drama of the Earth's history has been played and recorded. Not far, by Western Australian standards, from where the wasps built their nest and early humans painted self-portraits, lie the Jack Hills. The rocks of the Jack Hills have been around for three billion years and are too worn to form imposing topography. But in the hills are ridges of up-turned slabs of red and orange rocks, stacked against one another like old tombstones along a churchyard wall. They are sandstone and conglomerate (a poorly sorted sediment with pebbles as well as sand) that bear the hallmarks of being deposited by rivers three billion years ago. The majority of the sand grains are quartz. However, other durable minerals are present as well, including zircon, the December birthstone, a silicate of the element zirconium. Zircon crystals are typically small (fine or very fine sand size) but ubiquitous in minute quantities in igneous rocks. They too are survivors, and because they contain atoms of radioactive elements such as uranium, they are excellent timekeepers as the uranium decays. The rocks of the Jack Hills contain zircons (which themselves contain minute diamonds) brought long distances from their place of birth by the rivers that deposited the sands and conglomerates.

The Jack Hills zircon grains, collected patiently by crushing and sorting kilograms of rock to extract a thimbleful, include ones that are 4.4 billion years old, the oldest bits and pieces on Earth. This is important enough, but when they began to tell stories of their parentage, things became really extraordinary. A typically international collaboration of Australian, North American, Chinese, and European geologists has probed the atomic character of the grains, first looking at the details of the oxygen atoms. The most abundant form of oxygen on the Earth is oxygen-16, so named because each atom contains eight protons and eight neutrons; however, this form has a very rare sibling, oxygen-18, with ten neutrons, an isotope that is also stable and behaves like oxygen. The proportions of the two forms of oxygen in a mineral vary depending on the conditions under which the mineral formed. Igneous rocks that originate deep within the Earth have a distinctive and uniform value of this ratio. Rocks that originate at cooler temperatures, interacting with surface waters of rain and the oceans, have a very different ratio. Even if these cool rocks are remelted, a common fate in the early Earth, they retain that distinctive characteristic. And the Jack Hills zircon grains showed exactly that ratio—they testified to an origin inescapably associated with water. This contradicted all conventional wisdom, which assumed that no significant quantities of water would have existed for another several hundred million years—that is to say, this evidence, gathered from a thimbleful of sand grains, turned our understanding of the early Earth on its head. And there is more. As the atomic scrutiny continued, results showed that granites were the parents of the zircons, and granites only form the crust of continents. The diamonds within the zircons must have formed deep within a solid continental crust. Not only did the very early Earth have plenty of water, but, rather than being a churning molten hell, it also had *continents*.

Continents created by the raven in his marriage quest? The stories that individual sand grains can tell are endless, and we shall hear more tales later. William Blake and Robert Service had a point.

THE SCALE AND IMAGERY OF VERY SMALL THINGS

Sand grains are small, and they have come to symbolize the microscopic, to register the size of very small things in our imagination. Van Leeuwenhoek not only examined sand with his microscope but used a sand grain as a standard measure of his other discoveries. He described foraminifera as "little cockles . . . no bigger than a coarse sand-grain" and estimated the number of bacteria (which he called *animalcules*) that would fit in a grain of sand—more than a million of them. We have continued to use a grain of sand as a marker to help us grasp the scale of the small and the extremely large. The molecules that today's nanotechnology manipulates are a million times smaller than a grain of sand; one could fit as many nanoparticles into a grain of sand as one could fit grains of sand into a 1-kilometer (⅗ mi) cube. The *absolute* number doesn't matter; it's the *relative* scale, the imagery, that works. Each image captured by the Hubble telescope is of an area of space equivalent to that covered by a grain of sand held at arm's length. If the Earth were a grain of fine sand, the Sun would be 11 centimeters (4.3 in) in diameter and 11.5 meters (38 ft) distant; the closest star to the Sun would be 3,000 kilometers (1,900 mi) away.

Playwrights, poets, songwriters, and philosophers have all exploited the potential of a single grain of sand to resonate in the human imagination. As we have seen, we are able to extract extraordinary information from a sand grain, but not as much as Gottfried Leibniz thought possible: aspiring to greater horizons than Blake, he felt that the entire universe was there within a grain for our understanding. For Leibniz, a philosophical alchemist as well as a mathematician, scale was fundamental, and comprehending how small things accumulate to create the vast and, by the same token, themselves contain the infinite, was an ultimate goal. He proposed that the noise made by a single grain of sand moving with the waves is one of a series of tiny perceptions that we accumulate to hear the roar of the ocean. This image of the sound made by a single grain of sand is echoed by Samuel Beckett in *Waiting for Godot,* where Estragon and Vladimir discuss "all the dead voices": "They make a noise like wings." "Like leaves." "Like sand." The imagery of sand was used extensively by Beckett—it forms the contents of Lucky's suitcase in *Waiting for Godot,* and Winnie is buried in sand in *Happy Days.* Interviewed in his late seventies, Beckett reflected: "With diminished concentration, loss of memory, obscured intelligence . . . the more chance there is for saying something closest to what one really is. Even though everything seems inexpressible, there remains the need to express. A child needs to make a sand castle even though it makes no sense. In old age, with only a few grains of sand one has the greatest possibility" (Lawrence Shainberg, *The Paris Review,* Fall 1987).

A grain of sand has a significance and effect far beyond its size: "It isn't the mountain ahead that wears you out—it's the grain of sand in your shoe" (attributed to Robert Service). In his journals, John Cheever wrote of a grain of sand in the heart as the origin of self-destruction; Eddy Arnold sang of the life-changing capability of "one grain of sand." A single sand grain can have an influence far out of proportion to its size, but when it gathers together with vast numbers of its colleagues, very strange things indeed can happen.

Tribes

The Strange World of Granular Materials

History is a child building a sand castle by the sea, and that
child is the whole majesty of man's power in the world.

Heraclitus, ca. 535–475 B.C.

And so castles made of sand fall into the sea, eventually.

Jimi Hendrix

PLAYING WITH SAND

Watch kids digging at the beach, their concentration and intimacy with their
material. Scooping, scraping, molding, patting, sculpting, repairing, they are lost
in the cycles of excavation, construction, and destruction. Their hands are adapted
to different purposes, working both industrially and artistically. It's not just children
who enjoy playing with sand. Sitting on the beach, we adults also make patterns,
furrows, and heaps, or sift it through our fingers spontaneously, subconsciously, as
we talk or watch the kids. Sand, whether in its dry or wet mood, is a compelling
substance, somehow seductive and sensuous. Its topography conforms warmly to
ours; it flows through our hands, yet when we stand up, we can walk on it. There
is something *therapeutic* about sand, and indeed it has long played a role in healing ceremonies of widely different cultures. Today, sandplay therapy is a common
tool of psychiatry for both adults and children.

Playing with sand can be serious. Make a complex sand castle or simply a pile
on the beach, and you are conducting experiments that reveal some of the fundamentals of nature's character and materials in our everyday lives. Engineers,
physicists, space scientists, chemists, mathematicians, and biologists all play very

seriously with sand, and the results are often of considerable commercial importance. Why?

SOCIAL INTERACTIONS

Sand grains are gregarious, gathering together on a small or a gigantic scale, in your shoes or in desert dunes. When they congregate into communities, however, they exhibit strange social behaviors and interactions, often surprising, sometimes definitely counterintuitive, and always interesting.

Sand is a *granular material,* like rice, sugar, salt, coffee beans, the nuts in your cocktail assortment, cereal in its box, grain in storage silos, tablets in a medicine container and the contents of capsules, lawn fertilizer in your spreader, and countless other components of our daily lives. Granular materials behave very strangely. They are not really solids, they are not really liquids, but in many ways they behave more like the latter. Sometimes they behave more like gases, and some physicists believe they should be regarded as a type of matter all of their own. There are more questions than answers about their behavior. For example, *why* does sand form dunes of all shapes and sizes? Why doesn't it just spread itself around evenly? Watch the beach as the waves and tides move in and out. Why does the sand sometimes form itself into an undulating surface like a rippled cloth? Ralph Bagnold had a behavioral definition for sand that was based on a peculiar characteristic not shared by coarser or finer materials—the power of self-accumulation, sand's skill in using the energy of the wind to collect all its scattered grains and build them into heaps, separated by areas free of sand. In many areas of desert, for example, the sand is piled into dunes with bare rock in between, as if some giant combination of a vacuum cleaner and a bulldozer is constantly at work. We shall look more closely at this phenomenon in chapter 6.

Granular materials are everywhere in our everyday lives, and our lives often depend on how they behave. More than a billion tons of granular materials of one kind or another are produced annually in the United States alone. Pharmaceutical products often rely for their effectiveness on the proper mixing of granular components. In agriculture, mining, construction, and other industries, safely storing and working with granular materials in mountainous piles and heaps is vital. In

North America each year, more than one thousand silos containing granular materials collapse—suddenly, spontaneously, and for no apparent reason. Landslides, after hurtling down a mountainside, will often turn themselves into a kind of dry fluid, flowing out over level ground for long distances. We simply do not understand how and why they do this, but it is urgent that we learn. It is estimated that natural landslides cause a minimum of $1.5 billion worth of damage and kill at least twenty-five people each year in the United States.

Reflecting its potential fluidity and fickleness, sand, as the quintessential granular material, has become a symbol of instability and impermanence. The biblical admonition against building a house on sand may be exaggerated (see chapter 9), but the imagery is powerful. Albert Einstein described the left-hand side of the equation for his general theory of relativity as built on granite, the right-hand side on sand. Think of the symbolism of "ropes of sand" and "written in the sand"— and then there's Jimi Hendrix.

Understanding granular materials is vital to countless aspects of our lives, from making good cement to building sand castles, from ensuring that our medicines work to preventing breakfast cereals from jamming up in the box, from constructing sand traps on the golf course to laying the foundations of buildings. It plays a key role in military operations too: the black sands of Iwo Jima, made up of the ash from the island's volcano, have treacherous granular properties, one Marine describing the sand as "so soft it was like trying to run in loose coffee grounds."

SAND PILES

The next time you pile up sand on the beach, consider, however briefly, two things. First, the sorites paradox, a philosophical problem originated by Eubilides of Miletus in the fourth century B.C. Eubilides loved paradoxes (the classic "liar's paradox" was his), and perhaps he loved the beach, or at least playing with sand. For he became concerned about what is, and what is not, a "heap" (*sorites* derives from the Greek for "heap"). If a single grain of sand is removed from a heap, it clearly remains a heap, as it does after the removal of the next grain. So, asked Eubilides, when does it become a non-heap? The reverse is the same problem—a single sand grain is obviously not a heap, but when does a non-heap become a heap? The problem, philosophically, is one of *vagueness* and the imprecision of word usage. Though you might not be able to solve the philosophical problem, you can learn a great deal about the physics of nature by experimenting with making avalanches down the side of your heap. As you add grains to the pile, causing periodic avalanches, large and small, which look for all the world like flowing liquid, you are opening a window on how nature works. Sand piles are of intense interest to physicists.

However hard you try, there are clear limitations to what you can do with a pile of dry sand (or any other granular material). You can build it up until the sides reach a certain steepness and then, however many grains you add, it retains the same slope; a couple or a handful of grains will spontaneously tumble down the slope to maintain the same steepness. Dunes are gigantic piles of sand, constantly moving by avalanching (Figure 6). The physics of how sand piles behave was introduced by Bagnold, but their true weirdness and fundamental importance began to be demonstrated in the last years of Bagnold's life by a Danish physicist, Per Bak, and his colleagues. Their pioneering work was done at the Brookhaven National Laboratory on Long Island, New York, itself a great pile of sand.

Bak ultimately experimented with sand mathematically, but you can do on your kitchen floor the kinds of experiments that he started with. Take a variety of ingredients that are granular materials of different characters. Try using normal table salt (it may declare "fine flowing" or "it never rains but it pours" on the package), granulated sugar, assuming it is somewhat coarser-grained than the salt, and "sea salt," the big grains of angular salt to be put in a mill. On three different plates, pour out a good-sized pile of each. The three piles will have quite different appearances.

The granulated sugar makes a steeper pile than the fine salt, and the coarse sea salt is steeper still, with a more rounded peak. The steepness of each pile is a reflection of the material's natural *angle of repose,* the maximum slope at which the grains are stable. Try to steepen the slope and grains will tumble down the flank of the pile, reestablishing the angle of repose. The angle of repose is shallower, the slope gentler, for finer and rounder grains. Conversely, it's steeper for larger and more angular grains—the sea salt shows this clearly. The balance of the frictional relationships between individual grains versus the pull of gravity governs the angle of repose.

Salt and sugar aren't common sand, but they are granular materials, and all gran-

FIGURE 6. An avalanching dune in Egypt's Western Desert. Sand blowing over the crest is causing avalanches to sweep across the face of the dune; the rippled surface is caused by "liquid" sand in motion as the photograph was taken. (Photo by author)

ular materials behave in the same way. A pile of real sand may show an angle of repose steeper or shallower than fine salt, depending on the size of the grains and how round they are. To emulate the work of Bak, take a spoonful of sugar and drop a few grains at a time onto the pile, watching what happens. Some grains will roll down the side to the bottom of the pile, but not all. Some grains will set off miniature avalanches of sugar grains down the slope; some of these avalanches will continue to the bottom, while some will come to a halt on the way. But the angle of repose is maintained. However, it's next to impossible to predict what behavior will result from your next addition of grains.

Strictly speaking, there are two key angles for a pile of any particular material: the *angle of repose* is its slope after avalanching and coming to rest; the *angle of stability* is the slightly steeper slope that it can adopt as grains are added before avalanching. To keep things simple here, the former alone captures the principle of what is going on. The angle of repose of a given granular material is the key to its stability, and it is therefore of fundamental importance to us. A wall of sand seeking its angle of repose can create human tragedy: around the world, significant numbers of people, generally children and young adults, often at the beach, are killed by the collapse of excavations in the sand.

In detail, the slope of a pile of grains is irregular, the *average* being the angle of repose. As grains are added, the irregularities change their shape by the sliding of one or a few or many grains. The irregularities are, in a way, places of marginal stability, poised for instant change. Bak demonstrated in detail what you could see with your pile of sugar: the addition of a single grain anywhere on the heap can set off an avalanche of any magnitude. So is this whole process truly unpredictable? Yes and no. Yes, because predicting what will happen with each grain added is indeed impossible. No, because the general behavior of the pile over time *can* be predicted. Truly massive avalanches, those that affect large parts of the pile, are very rare. Far more common are small avalanches, and the smaller they are, the more frequent. If you plot a graph of the frequency of avalanches involving different numbers of grains, it will be overwhelmed by the huge number of very small events. But analyze the numbers a little differently and some remarkable characteristics show up, a pattern, an order to things—nature at work. To see the importance and applications of this work, we need to take a moment for a little mathematics.

PLOT LINES

Statistics has a long history as an important but somewhat opaque and difficult area of endeavor. From the early data collection on the range of height of individuals in a population, any number of successful efforts have been made to squeeze order out of what seems to be natural randomness. The "bell-shaped curve" has become a standard, qualitative expression of this. In any population of natural occurrences, whether humans, sand grains, or tosses of a pair of dice, there is a peak, the most common occurrence of whatever is being measured—height, diameter, or added numbers. The peak is always around the middle of the range, with fewer and fewer of successively bigger or smaller measurements. If these kinds of results,

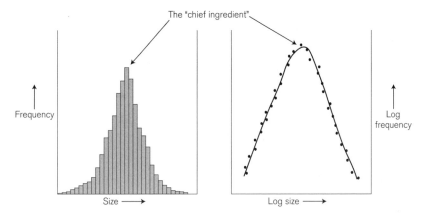

FIGURE 7. Sand size distributions, illustrating Johan Udden's "law of the chief ingredient": a "normal" histogram (left) and the same data plotted logarithmically (right).

say people's heights, are plotted on a graph that shows the heights set out in group increments against the number of people in each group, then the graph looks like a bell: at the top is the most frequently occurring height, which is flanked by the successively decreasing frequencies of taller or shorter people. This is the so-called normal distribution, because it crops up so commonly that it seems to be nature's way of organizing things. And it's not at all random; it has a very clear shape to it.

This is true of sand size distributions, as Johan Udden noted with his "law of the chief ingredient." For a sample of sand from a particular place, a graph of the frequency of grain size occurrences against categories of size often looks like a bell curve, a normal distribution (Figure 7, left). For dune sands, it is a tall and narrow bell—they are well sorted. For river sands, it is a squat bell, showing the great range in size.

Bagnold, however, recognized that the bell curve might not be the most useful visual representation of sand size distribution. The bell curve displays size for the sake of size, the very smallest and the very largest grains disappearing off into the indiscernible extremes of the curve. But those ends of the population have a lot to tell us. The Udden-Wentworth grain size scale was developed on the basis of *ratios* or *multiples* of size, giving very small grains equal time with the big boys. Better, surely, to honor the principle of the grain size scale and show size frequencies not

in terms of a linear scale but of a ratio scale? This was another of Bagnold's pioneering approaches, plotting graphs of sizes and frequencies on *logarithmic* rather
than linear scales.

We don't need to enter into a deep mathematical discussion, particularly about
logarithms, but they are critical in discerning what is going on with sand sizes—
and piles—so a brief excursion is necessary. All Bagnold was doing by using a logarithmic scale was applying the principle of the importance of ratios or multiples.
Each of the ascending major categories on the grain size scale is twice as large as
the grain size of the previous category (Figure 1). Plotted on a logarithmic scale,
the *length of each of these increments is the same.* When used to plot a graph of grain
size frequencies, a logarithmic scale stretches out the portrayal of grain sizes to give
every size range an equal say, and allows us to distinguish as easily among very small
grains as among very large.

In order to display the characteristics of his samples, Bagnold not only plotted
size in multiples (logarithms) but also plotted frequencies that way. This way, the
curving sides of the bell curve graph became *dramatically straight lines* (Figure 7,
right) and the tail ends of the distribution were no longer lost. This was not the first
time this had been done, but his work was pioneering in gathering huge volumes
of data, developing the details, and interpreting the results. The character of his
plots, in particular that of the straight lines, proved to be invaluable in comparing
and contrasting different communities of sand grains.

Interpreting the details of these kinds of graphs can quickly become another arcane science. But the principles behind them have a wide-ranging significance.
Through detailed statistical analysis (not to be reproduced here), Bagnold argued
for the fundamental importance of this kind of distribution in our world. And in
thinking this way, that "Nature is concerned with relative rather than absolute magnitudes," he was among those who anticipated some imminent and extraordinary
revelations of physics: the physics, as Bak demonstrated, symbolically enough, of
sand piles and their avalanches.

A conventional graph of the frequency of avalanches involving different numbers of grains is overwhelmed by the huge number of very small events, and it seems
random, characterless. But bearing in mind nature's concern with relative magni-

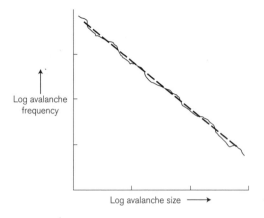

Log avalanche
frequency

Log avalanche size ⟶

FIGURE 8. The size of sand pile avalanches versus their frequency, plotted logarithmically, is a straight line.

tudes, it is time to use logarithmic scales again. Bak showed that plotting his avalanches this way resulted in a straight line (Figure 8).

Straight lines in Bagnold's graphs of sand sizes, straight lines in Bak's graphs of sand avalanches—what is going on? If a complex natural phenomenon becomes simple when displayed using logarithms, it is proof of nature's liking for multiplication rather than addition. Lots and lots of small things and only a few big things, whether they are sand grains or earthquakes, is the rule in natural processes, a fundamental way in which nature operates. (Per Bak's book discussing the behavior of sand piles is titled *How Nature Works.*) The laws governing such processes are referred to as *scaling* or *power* laws: the frequency (or its logarithm) of occurrence of something is proportional to a measure (often size) of that something *multiplied* by itself a specific number of times (raised to a certain power, mathematically speaking). Newton's law of gravity is a power law; the pull of gravity on an object decreases with distance to the object *squared.* Since it decreases, Newton's law is an *inverse* power law—and so is that of sand avalanches: the bigger the event, the more rare it is.

But what about the "real world"? Scaling laws show up everywhere—in earthquake magnitudes (each successively larger magnitude on the scale is a multiple of the previous), population distributions, city sizes, the brightness of the Sun, and music (the structure of rock music, classical music, and the spoken word all obey scaling laws). And the real world of sand and other granular materials is, unsurprisingly, full of examples. Bak was more interested in the underlying mathematics than with messy reality; he wrote in *How Nature Works* that "the experiments on sand turned out to be much more complicated and tedious than we had anticipated." But then, "Don't get me wrong. I have the deepest respect for the type of science where you put on your rubber boots and walk out into the field to collect data about specific events." Fortunately, many scientists, particularly geologists, have been willing to do exactly that. They have amassed a substantial body of data on landslides in the Himalayas, documenting that the slides range in size from a few wheelbarrowfuls to ten thousand dump truck loads: they obey a power law. Flushes of sand and mud down into the deep ocean also follow a power law in their frequency versus size. And, if you are on a beach with sand dunes behind it, look for the avalanches down the sand slopes or make some yourself: they are obeying power laws.

The more you look, the more interesting sand avalanches and their laws become. As sand grains are added to a pile and avalanches maintain the slope, it seems as if the system, the pile, is *poised* in a constant state of almost instability. Bak termed this *self-organized criticality,* a state in which a very small event can trigger huge consequences—or very small ones. A pile of sand is a self-organized critical system, disturbed by single grains moving down its side or large-scale avalanches—it is a simple model of many seemingly complex natural systems.

Sand piles have even stranger characteristics, some of which may explain why grain silos spontaneously collapse. You would think that the greatest weight of a sand pile, and therefore the greatest pressure exerted at its base, would be directly under the tallest part of the pile, below the greatest weight of sand. But this is not always the case. The weight and pressure distribution within a pile of any granular material is determined by the way in which the individual grains contact each other and distribute the stress. Quite commonly, grain shapes and sizes mean that there are microscopic chains and networks of grains that are oriented and in contact with each other in such a way that they carry most of the pressure from the weight of the material above them. These chains seem to behave like the soaring arches of Gothic cathedrals, which serve to transmit the weight of the roof, perhaps a great dome, outward to the walls, which bear the load. In a sand pile, particularly one that is confined in a container of some sort, these chains perform the same function—they carry the stress outward to the container, rather than directly downward to the base of the pile. If grains of wheat or rice in a silo organize themselves in this way, the resulting stress may cause sudden and catastrophic failure of the structure if it's not designed to withstand it. It's this same mechanism that allows sand-filled timers, or hourglasses, to work so well (see chapter 9) and helps support your weight at the beach.

The behavior of sand, or any granular material, is strange enough even when the grains are more or less uniform and the only external influence is dropping an additional grain or two onto a pile. When the grains are different sizes and different densities, and when they are poured or shaken a little, things become seriously bizarre.

SHAKING AND STIRRING, MIXING AND UNMIXING

In Russian folklore, there are many stories of Baba Yaga, a terrifying ogress who eats children and flies around in a mortar using the pestle to steer. One of the tales stars Vasalissa the Beautiful, a merchant's only daughter who is sometimes, for reasons that will become obvious, referred to as "the Russian Cinderella." On her deathbed, Vasalissa's mother gave her a doll that would look after her, and when the father remarried—to, yes, a cruel stepmother with two ugly daughters—the doll helped Vasalissa complete all the menial tasks that were forced on her. One day, the two sisters dispatched Vasalissa to fetch fire from Baba Yaga's house (a log cabin that moves around the forest on chicken legs), but the girl was kept captive by the ogress and required to perform tasks in return for fire. In addition to the inevitable domestic chores, Vasalissa was given a number of seemingly impossible tasks, one of which was to pick out grains of flour from a bucket of sand (the ingredients vary in different versions of the tale, but it is always a granular materials problem). Fortunately, Vasalissa had her doll to help her, and the job, with great combined effort, was done. She eventually escaped with the fire, the stepmother and the ugly sisters burned to death, and Vasalissa married the tsar.

But had Vasalissa (or her doll) known a little of the physics of granular materials, she could have accomplished the task of separating the flour from the sand with ease. Some years ago, when the mysteries of the physics of granular materials were beginning to be revealed, a group led by Hernán Makse at Boston University demonstrated that a mixture of sand and sugar would separate into its components if you

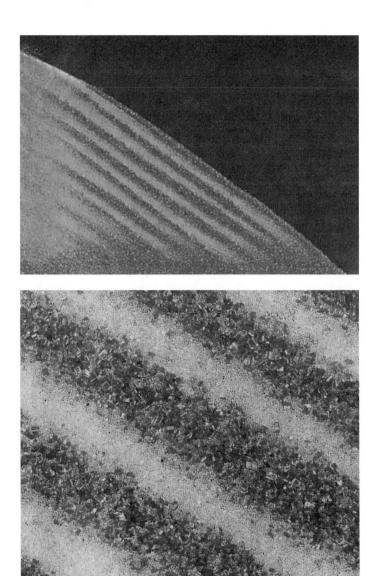

FIGURE 9. Segregation of small glass beads and larger sugar grains (dark) as a result of the simple act of pouring; below, a detailed view. (Photo reprinted by permission of Hernán Makse and Macmillan Publishers Ltd. © 1997 Nature Publishing Group)

simply poured it slowly and evenly into a pile. "A miracle occurs," said Eugene Stanley, one of the collaborators. "It's like throwing a deck of cards on the table and having all the blacks fall on one side and the reds on the other." Makse, Stanley, and their colleagues conducted a series of experiments using different materials, including sand, glass beads, and sugar crystals, and the results were clear. Typically, larger grains will have a steeper angle of repose than smaller ones, and they will roll down the slope more energetically. The smaller grains tend to get stuck at the top of the pile, the larger ones at the base—they spontaneously segregate (Figure 9). But things become more complicated. As the different angles of repose of different grains are reached and exceeded, successive avalanches will be made up of different-sized grains. The cascades of smaller grains will stop first, to be then covered by a layer of the larger grains still on the move. The process repeats itself over and over, creating a layered pile.

Different (and unpredictable) results can be achieved by varying the size, density, and shape of the grains (and therefore their angle of stability or repose), but spontaneous segregation is a common phenomenon. As we shall see, all natural sands and sandstones are layered in one way or another, regardless of whether they were formed underwater or by wind, and clearly understanding the kinds of spontaneous segregation and layering demonstrated in Makse's experiments is fundamental to understanding how these processes work. Figure 9 is an experimental version of what is happening in the sand dune avalanches of Figure 6. Furthermore, pouring of granular materials is a common industrial activity, and this research begins to hint at how critical an understanding of these strange behaviors is to ensuring that undesirable things don't happen in the process.

But pouring is only the start of the story. I travel with a container of assorted vitamins and other pills. If I want one of the small items—say, a daily aspirin—it becomes quite frustrating because all the small pills have settled at the bottom of the container and all the larger ones at the top. The same thing occurs with a jar of assorted cocktail nuts—open it up, and the contents have sorted themselves out, with the Brazil nuts all at the top (particularly irritating to me, since I am violently allergic to them). Cocktail nuts have lent their name to a phenomenon that has come to symbolize the strange sorting behavior of granular materials: the Brazil nut effect. The same thing can be observed on opening a box of breakfast cereal where the components are different sizes or weights—the big bits are all at the top. The effect was first documented in the 1930s, but explanation was a challenge.

Today, research into granular behavior takes place at universities all over the world, but in the 1990s it was Sidney Nagel and his colleagues at the University of Chicago who brought the science to the world's attention when they demonstrated some of the bizarre origins of the Brazil nut effect. In a 1999 *Scientific American Frontiers* program on PBS, it was suggested by the interviewer, Alan Alda, that Nagel really loved sand. His response was, "Of course. It's one of the best substances there is." Sand, glass beads, sugar, salt—it doesn't matter what the granular material is, Nagel loves playing with it. To examine the Brazil nut effect, he buried a large glass bead (the Brazil nut) in a jar of small beads (the peanuts) and shook it. After each shake, students were required to meticulously pick out each bead in order to see what was going on; Heinrich Jaeger, who is now a leading researcher at Chicago, was one of those students and described it as a painful experiment. But it was worthwhile. While previous explanations had suggested that it was simply a matter of small things filling in the holes under larger things (percolation), Nagel's work demonstrated that *convection* was occurring—convection, as in a genuine liquid, where hot liquid rises, cools, and falls, setting up circulating cells—but this convection was happening in a jar of glass beads. As shaking occurs, the beads at the center move upward, but the larger ones, the Brazil nuts, become marooned at the top because the downward movement takes place along the sides of the container and they are too big to participate. Today, techniques of medical imaging have been borrowed that, combined with time-lapse photography, save tedious hours of student labor and enable detailed documentation of convection in granular materials. Bob Hartley and Bob Behringer, at Duke University, have spent a lot of time vibrating piles of sand and filming their behavior. Figure 10 dramatically illustrates convection patterns in one of their piles: it *looks* like a liquid.

These are startling, gravity-defying results, but they don't mean that we understand fundamentally what is going on. This would require the ability to model and construct the mathematics for the behavior of each individual grain, which even with today's computing power is impractical. The Brazil nut effect may well result from a conspiracy of percolation and convection. Mysteries remain, but at least we can experiment with the strange behaviors of granular materials and apply this

FIGURE 10. Convection in a vibrating sand pile. This view is of a vertical slice through the pile, seen from the side. Some grains have been colored (dark) to highlight the convection pattern. (Photo courtesy of R. R. Hartley and R. P. Behringer, Duke University)

knowledge when we need to mix efficiently or transport safely such things as breakfast cereals, medicines, coal, or fertilizer. And do these behaviors account for the annual crop of stones and boulders that appear on the surface of farmers' fields each spring? Or is that simply the result of freezing and thawing? Fragments of meteorites that may have fallen thousands of years ago remain on the surface of the great Saharan dunes—extraterrestrial Brazil nuts? And while, as we saw in the previous chapter, many modern and ancient sands show graded bedding, where the large grains are deposited first, the smaller ones later as the current wanes, occasionally we find *reverse* grading. Layers where the largest grains are on the top are typical of rapid and violent depositional events (chapter 5). Sand-covered slopes in the shallow ocean may become unstable, perhaps shaken by an earthquake, and the sand and mud hurtle down into deeper water as debris flows; when they settle out, the larger grains are often at the top—deep-ocean Brazil nuts?

The story continues. You will by now be sufficiently familiar with the strange behavior of granular materials to expect the unexpected, so the fact that a "reverse Brazil nut effect," the RBN, has been documented will not come as a surprise: large *lightweight* objects can, counterintuitively, sink through a bed of smaller, denser grains. A whole range of behaviors are sensitively dependent on densities of the grains, speed of vibration, and other variables.

The more research plays with sand, the more complex granular materials appear to be. Grains of different sizes and materials really seem, given the slightest provocation, to prefer the company of their own kind, separating themselves from the others. It seems easier to unmix things than to mix them. Put well-mixed rice and split peas into a horizontal glass drum, start the drum rotating slowly, and go away for a couple of hours; when you come back, the components will have organized themselves into alternating bands of gregarious peas and socially bonding rice along the length of the container. Repeat the experiment with other materials and look lengthwise along the axis of the drum: there will be one of any number of exotic geometrical patterns of self-segregated granular materials.

Why does this happen? Troy Shinbrot, a researcher at Rutgers University who has worked on segregation in drums and the RBN, often in collaboration with Jaeger, has written: "The RBN is sure to provide fruit for further exploration and debate. We find ourselves facing the situation anticipated by Mark Twain: 'The researches of many commentators have already thrown much darkness on this subject, and it is probable that, if they continue, we shall soon know nothing at all about it.' Although farmers can count on continuing harvests of heavy boulders under any segregation model, it remains to be seen whether—and how—pharmaceutical engineers should expect their granular formulations to mix or separate."

MUSIC OF THE SPHERES

If you play in a rock band, try putting a jar filled with a mixture of granular materials from the kitchen in front of the bass amplifier during a gig. Strange patterns may appear, or components may separate in response to the musical frequencies— it's happened in laboratories. By the same token, nothing at all may happen.

In the late eighteenth century, a lawyer, musician, and scientist in Leipzig, Ernst

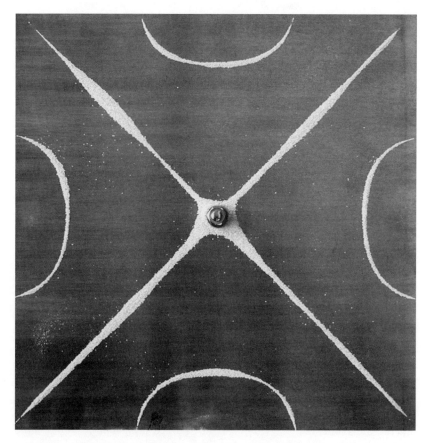

FIGURE 11. A Chladni pattern. (Photo courtesy of Paul Koza, Department of Physics, Ludwig-Maximilians-University, Munich)

Chladni, was determined to make sound waves visible. He succeeded in doing so by covering glass or metal plates with sand and drawing a violin bow across the plates' edges. The vibrations provoked the sand grains into a frenetic, leaping dance, not just randomly: they arranged themselves like Scottish dancing groups into formations. The patterns were highly variable—stars with different numbers of points, crosses, complex intersecting arcs (Figure 11). The science of acoustics and the physical manifestation of sound was born.

Over the years, Chladni's experiments generated increasingly complex patterns, their shapes animated according to frequency and amplitude. In the nineteenth century, Michael Faraday, the British scientist who revealed the fundamentals of electricity and magnetism, first observed convection in granular materials (although not, presumably, in a jar of cocktail nuts). Faraday also demonstrated that vibrating a fluid generates complex standing wave patterns on its surface. Similar to Chladni's patterns, these were referred to as *Faraday waves* or, more poetically, *crispations.* Once again, the similarity in behavior between granular materials and real liquids is disturbingly apparent.

The patterns produced in Chladni's experiments were as much about the material on which they were produced as about the sand itself, but the patterns seem to relate to underlying natural behaviors. All granular materials indulge in some extraordinary pattern making. Faraday's crispations appear on the surface of a container of vibrated sand, just as they do in a conventional liquid. Heinrich Jaeger and Troy Shinbrot have worked on this phenomenon too, together with researchers elsewhere, such as Paul Umbanhowar, documenting standing waves in a vast variety of stripes, squares, hexagons, and interlocking, fractal-like patterns— not to mention structures that resemble the stitching on a baseball (Figure 12, left).

The patterns, as Chladni found, respond dramatically to changes in frequency and amplitude, but this relationship is complex. Umbanhowar and others at the University of Texas first demonstrated, accidentally, what happens at low frequencies: clumps of sand grains bounce up and down, alternately leaping into spouts and subsiding into hollows (Figure 12, right). Termed *oscillons,* these clumps move slowly around the surface of the sand, sometimes combining with and sometimes repelling others, but generally remaining stable for long periods of time. This is yet another example of the bizarre behavior of granular materials. Considerations of particle interactions, energy transfer, impacts, and other aspects of classical physics shed some light on these patterns, but what exactly is going on is not entirely clear. The patterns' significance goes further than aesthetics and oddball behaviors, however: there is a startling resemblance to patterns in atomic lattices, quantum physics, the weather, and a host of other natural forms. The team at Chicago led by Nagel and Jaeger has recently reported that a jet of sand fired at a small circular target mimics structures interpreted as characteristic of the birth of the universe.

FIGURE 12. Patterns formed on the surface of vibrated sand, viewed from above (left, top and bottom); an oscillon in action, with metal beads mimicking very fine sand grains (right). The four dishes at top left are each 12.6 cm (5 in) across; the hexagons in the image below are each approximately 0.6 cm (¼ in) across, and the oscillon is 2.7 millimeters (¹⁄₁₀ in) across at its base. (Photos courtesy of Paul Umbanhowar)

In November 2000, Jaeger, Shinbrot, and Umbanhowar published a short article in *Proceedings of the National Academy of Sciences,* titled "Does the Granular Matter?" Their answer is hardly surprising. It is quite possible that dancing, as well as avalanching, sand grains will help us understand some of the fundamental ways in which nature works.

Much of sand's behavior is a result of the spaces in between the grains and what those spaces contain, as well as of the grains themselves. The air in dry sand contributes to avalanching and the Brazil nut effect—if the same experiments are done

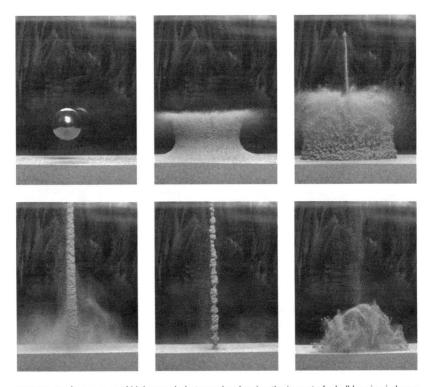

FIGURE 13. A sequence of high-speed photographs showing the impact of a ball bearing in loose sand causing the eruption of a "grainy geyser." (Photos courtesy of Detlef Lohse and the Physics of Fluids Group, University of Twente, The Netherlands; see Lohse et al., *Phys. Rev. Lett.* 93, article 198003 [2004])

in a vacuum, the results are not always the same—and fundamentally affects its properties. Golfers should be especially concerned about the air in the spaces between sand grains and may find the experiment shown in Figure 13 somewhat distressing. Physicists at the University of Twente in the Netherlands (and Jaeger's group, again) simply dropped a ball bearing into loose sand and generated a new kind of matter—a fluid of dense material and air. The air driven into the sand by the impact, together with the air between the grains, is compressed and erupts, forming a delicate geyser that, very briefly, reaches higher than the level from which the ball was dropped.

For a golfer, the lie of the ball would be soul-destroying, but this is why a great deal of attention (and science) is devoted to the character and properties of the sand in bunkers and sand traps. One of those properties, the factors that determine the all-important lie and "playability" of an errant golf ball on landing in sand, is the firmness—how well packed together the sand is. The size, sorting, and shapes of the grains are critical—the applied science of sand in golf course design is a whole discipline in its own right. And, of course, the sand trap plays very differently when it's wet.

So far, we have covered some of the extraordinary ways in which *dry* sand grains behave when they get together, but add a little water and everything changes. Dry sand itself behaves eerily like a liquid, but wet sand behaves more like a solid—as long as it's not too wet.

JUST ADD WATER–ANOTHER MIRACLE OCCURS

Why can sand castles only be built with wet sand? Why is *how wet* the sand is so important? Why do our footprints on the beach leave a pale, drier "halo" around them? Why is wet sand darker than dry sand? This last question is the simplest to answer—the multiple refractions caused by water in the sand cause light to spend more time bouncing around the sand, and less light escapes, so the sand looks darker. The other questions are a bit trickier. As Alan Alda remarked in his interview with Sidney Nagel, "The beach is a great place for thinking up trivial questions"—but the answers and their implications are far from trivial.

As every kid knows, you need wet sand to build a sand castle. Sand sculpture is an old art form and a global one. Today, festivals, exhibitions, competitions, corporate events, advertising, and therapy feature the art of sculpting sand into often huge—and very often beautiful—representational and abstract forms. The material that, when dry, runs between our fingers, avalanches, and blows away becomes robust and *workable* when it is wet.

Larry Nelson is a sand sculptor based in Venice Beach, California, and his work

FIGURE 14. Examples of Larry Nelson's sand sculptures. Each is a little over a meter (3.5 ft) in height. (Photos courtesy of the artist)

is extraordinary. He achieves forms with sand that seem impossible: fragile soaring arches, interlocking curves (Figure 14). To achieve these exquisite results, he has developed an intimate relationship with his material, and his description of the character of wet sand eloquently summarizes its magic:

It doesn't take long to learn that water mixed with sand changes both completely. Two fluids become a formable solid. It may be counterintuitive, but anyone who has spent time in the playground knows this in the fingers.

Water tends to pull in on itself. This makes raindrops, rainbows, and flying sparks from every sprinkler. Anything in contact with the water feels this pull, imparting a tiny tensile characteristic to damp sand. It sticks together.

This adhesion is the essence of sand sculpture. It's also the bane, because damp sand stickily resists being compacted beyond a certain point, no matter how hard it's hit. Granular materials naturally form arch structures, tiny but powerfully resistant to compaction. Smack it on top, and your pile simply spreads sideways.

The closer the sand grains, the better they stick together. The sand has to be wet, but how do you get it to pack well if the constituent grains hang up on each other, leaving relatively large air spaces? What you want is grains fitting closely together, filling space as densely as possible so that surface tension embraces as many as possible.

The answer is in the water. It is a lubricant. Not a very good one, but effective if there's enough of it to give the grains a little time, as they sink, to become neighborly. In the final milliseconds of their fall through the water, you give them some help by rapidly striking the pile with a stick. This shakes the whole assembly, helping the grains snuggle in next to each other, minimizing gaps, giving the water remaining after most has drained away the strongest possible handle on the sand.

As Nelson says, the answer is in the water. Surface tension is a force that operates at the interface between water and air, causing the slight bulge on the water in a full glass, pulling liquid slightly up a straw by capillary action, and allowing insects to walk on the surface of a pond. Surface tension is incredibly strong—the microscopic attraction between the capillary films of water coating two neighboring sand grains is greater than gravity; the grains are very effectively stuck together. Even with the addition of a very small amount of water, a network of sticky bridges is built up within the wet sand and its character changes completely; its angle of stability increases dramatically. You can illustrate the strength of surface tension by wetting two ping-pong balls and putting them on a table: touch them together and they stick, allowing you to pull one with the other across the surface.

We probably know less about the physics of wet sand than we do about its dry version, but the food and pharmaceutical industries, never mind landslide investigators, need the information. Only recently has it been appreciated how little water is needed to effect a dramatic transition (although sand sculptors have known this empirically for a long time). Sand becomes a miraculously solid material with the addition of as little as 1 percent water—and it preserves this character even when the water content is more than 10 percent. The key lies in the strength of surface tension, which depends on the boundary between the water and the air in the spaces between the grains. As more water is added, the total length of that boundary decreases and the adhesion between the grains diminishes. Eventually, you end up with a slurry (sand becomes a liquid again), and, of course, the total absence of that boundary is why you can't build a sand castle underwater. Unless, that is, you are using *magic* or *space* sand, which behaves in exactly the opposite way as normal sand: the grains have been specially coated and treated so as to be water-repellent (hydrophobic); this causes them to clump and cluster together underwater but become loose when brought into the air.

Nelson is an expert and unique sand sculptor. He devotes a lot of attention to the water he uses and the careful, continuous packing of the sand in order to be able to sculpt vertical and overhanging surfaces into his signature floating, hollow sculptures. Often he packs the sand inside a fabric or plastic form, which provides a containing pressure as the grains get themselves organized. An alternative technique is to build with layers of dripping wet sand, using no form. But as he says: "No matter how you make the pile, the truth comes out as soon as you start to carve. The nature of the pile comes from the sand of which it's made, the way you pack it, the shape of the grains, and the type of water you used. It takes time to know sand."

The way in which sand packs itself is an area of investigation in its own right. Even if all the grains are the same size and shaped as perfect spheres, there are still heated arguments about how packing works, in spite of the fact that scientists and mathematicians have been studying it for centuries. Put marbles in a jar and gently shake it: they will jam themselves into "random close packing," the marbles occupying around 64 percent of the total packed volume and air occupying the other 36 percent. To pack the marbles more efficiently, or more densely, the packing needs to be organized rather than random. Shake the jar vigorously so that the marbles bounce up and down a little, and they will settle into an ordered hexagonal arrangement exactly like oranges stacked at the vegetable store. They now occupy 74 percent of the volume. That this is the maximum density for spheres was first proposed by Johannes Kepler four hundred years ago, as he mapped the universe and sought its inner harmony and structure. But "Kepler's conjecture" took on a similar stature to "Fermat's last theorem" and was proved mathematically only in the closing years of the twentieth century. Continuing work shows that the state of packing achieved by any group of spheres that may be poured, stirred, or shaken depends very much on *how* they are poured, stirred, or shaken. Larry Nelson knows a lot about that, and he is handling grains that are not all the same size and not all perfect spheres. In his comment quoted above, about giving "the grains a little time, as they sink, to become neighborly," he's describing one of the processes that researchers are curious about. They are also worrying about another detail that helps sand sculptors—the tendency of grains in certain packing structures to jam up, to become jam-packed. Magicians have performed a dramatic trick based on this. The performer produces a pot of sand, into which he inserts a knife and withdraws it, repeating this several times to show how loose the sand is. Plunging the knife in one last time, the magician gives the pot a fairly vigorous shake and swings it around his head by the handle of the knife: the shaking caused the sand to enter into this peculiar jammed state, as everyday granular substances seem to do with irritating frequency.

QUICKSAND

But let's return to the beach. It is the conspiracy of grains with water and air in the spaces in between that weaves the magic—and explains why our footprints appear the way they do. For the magic mix has yet another extraordinary property. Sand settled through water below the high-tide line is extremely well organized and closely packed. When we walk on it, this structure is disturbed and our weight actually causes the spaces in between the grains underfoot to *expand*—allowing water to drain into the increased volume and creating the dry halos around our footprints. This effect is known as *Reynolds dilatancy*, after Osborne Reynolds, the nineteenth-century scientist who pioneered our understanding of fluid dynamics. Fill a balloon with well-packed sand and attach it to an open-ended glass tube with a piece of cloth over the balloon end so as to allow the movement of water but not sand. Hold this contraption upright with the open end of the tube uppermost; pour water into the tube until the sand in the balloon is saturated and water appears in the tube. Now squeeze the balloon—contrary to expectations, the water level in the tube *drops*. You have demonstrated Reynolds dilatancy and explained the footsteps-on-the-beach phenomenon.

But this habit of dilating is not simply an amusing and counterintuitive property of sand. It's extremely important—and dangerous. For a start, it causes quicksand. A hapless character being sucked in by quicksand is a movie classic: the character is perhaps saved, perhaps not, and a hat, a Stetson, or a pith helmet floating on the surface is all that remains. *The Hound of the Baskervilles, Ice Cold in Alex, Lawrence of Arabia,* and *The Jungle Book* are just a few movies that feature a dramatic—but completely unrealistic—quicksand scene. It's simple to demonstrate that it's impossible to sink completely in quicksand, never mind being "sucked in."

But it's also essentially impossible to pull someone out—there is a real risk that the victim will be torn apart. Quicksand forms when there is sufficient water in between the grains to separate them—to push them apart through dilatancy—but the water is prevented from draining; the sand is in suspension. This can happen when an incoming tide scours large holes in the sand that are rapidly filled by the outgoing tide, trapping water and air in the sand. Or a subsurface spring or other source of water percolates upward through a body of sand, dilating it. The result is a slurry, delicately balanced between solid and liquid, switching instantly but briefly between the two states with the slightest disturbance. But being a mix of water and sand, quicksand is more dense than water, and the human body floats well in it; this was demonstrated effectively in an episode of the highly entertaining television series *Mythbusters,* where the presenters bobbed around like corks in a huge tub of quicksand.

The problem arises when a person floating in quicksand tries to move too quickly; the movement destroys the dilatancy of the slurry and the grains reconvene and jam back into a solid, effectively cementing the unfortunate person in place. It has been estimated that the force needed to pull your foot out of jammed quicksand is about that needed to lift a medium-sized car. The key is to wiggle, allowing water to fill the space created around you, and then swim, very slowly. Quicksand is lethal because lone individuals die of exposure, starve, or drown when the tide comes in, not because they are sucked under.

Dry quicksand is something else altogether. Golfers will be relieved to know that the sand in Figure 13 was prepared by blowing air through it to create a very loosely packed bed, with the result that the ball bearing disappeared completely. There are tales, as in the stories of T. E. Lawrence, of camels and vehicles disappearing completely in dry desert quicksand, but it has never been documented as being lethal. (The dry quicksand scene in *Indiana Jones and the Kingdom of the Crystal Skull* is relatively realistic—at least Professor Jones has time to deliver a short lecture on the physics of their predicament.) It has been suggested that there is no such thing as dry quicksand, but that is also incorrect. Ralph Bagnold, in his autobiography, *Sand, Wind, and War,* observed in the Egyptian desert that "there were other places, pools of 'liquid' sand, in appearance just like the rest of the surface, but so soft one could plunge a six-foot rod vertically into them without effort." And ask Keith Tabscott. In July 2007, he lowered himself into a sand hopper at an industrial plant in Florida in order to try to free a blockage. The sand collapsed on him up to his chest, and the more he struggled, the more the sand locked around his legs. Fortunately, he was wearing a safety harness; news helicopters were kept away in case their vibrations caused further avalanching and settlement of the sand, and it took two hours to free him, unharmed. He was lucky: fatal accidents have occurred in sand hoppers. Perhaps this was not truly "dry quicksand," but don't tell Keith that.

SHAKING EARTH

Quicksand may be a problem, but it's a relatively small-scale one. Dilatancy can also cause problems on a devastating scale, when what is going on in the balloon experiment causes sand and soil to dilate and liquefy under houses, office blocks, and entire towns. Shake waterlogged sand, and liquefaction occurs: the grains move apart and the friction and adhesion between them is lost, destroying the stability of the material, which then flows and compacts, expelling the water. Earthquakes are particularly good at doing this, with catastrophic consequences. The earthquake that devastated the region around Bhuj in northwestern India in 2001 was one of the most damaging in the country's history. Twenty thousand people were killed and the havoc caused more than $3 billion of damage. Because the vibrations provoked the sand and soil into liquefying, buildings no longer had a foundation, causing them to sink, tilt, and collapse. Huge cracks in the earth appeared, and water and sand erupted in volcanic fountains as the ground collapsed back on itself. This is a common result of earthquakes: more damage is done by liquefaction than by the tremors themselves. This was the case in the great San Francisco earthquake of 1906, in the Anchorage event in 1964, and in the Loma Prieta earthquake in 1989, when some buildings in San Francisco sank until their third floors were at ground level, and sand volcanoes erupted from the ground (Plate 5). Liquefaction was also amply demonstrated in the most powerful series of earthquakes to strike the lower 48 in recorded history—not in California, but in New Madrid, Missouri, between 1811 and 1812.

New Madrid at that time was, fortunately, a small town of four hundred inhabitants, none of whom were aware that they lived above an ancient system of deep fractures in the Earth's crust beneath the Mississippi River valley. The fractures do not exercise themselves very often, but when they do it is with extreme violence. In the early hours of December 16, 1811, the terrified residents were awoken by violent shaking, accompanied by an appalling roaring sound. They later reported that the surface of the Earth moved in waves and that cracks opened in the ground from which water and sand erupted. One resident, Eliza Bryan, wrote that "the surface of hundreds of acres was, from time to time, covered over in various depths by the sand which issued from the fissures, which were made in great numbers all over this country, some of which closed up immediately after they had vomited forth their sand and water" (quoted in *Lorenzo Dow's Journal*, 1849).

We are often tempted, for dramatic purposes, to compare the destructive power of natural events such as earthquakes to the biblical demise of Sodom and Gomorrah. Ironically, work by David Neev of the Geological Survey of Israel and K. O. Emery of Woods Hole Oceanographic Institution, together with research by engineers and geologists in the United Kingdom, has suggested that, assuming the twin cities existed at all, they may well have been destroyed by earthquakes and liquefaction. The area around the Dead Sea, the likely location for Sodom and Gomorrah, lies across the boundary of two rapidly shifting segments of the Earth's crust and has experienced significant earthquakes over recorded history. Lying below sea level, it is also the ultimate destination of vast volumes of sand and mud, and in many of these ancient layers are the telltale signs of violent water expulsion from the sediments. Large-scale slumping is also evident. Altogether, these forensic clues point to an obvious culprit for the destruction. But what about the fire and brimstone? Bitumen, natural asphalt or tar, was much prized during ancient times for medicine, the caulking of boats, and the preservation of mummies, and sulfurous bitumen, together with lighter oil, has for thousands of years been leaking from fractures around the Dead Sea, seeping into the ground and the water. All that would be required would be gas leaking along with the oil and a spark as the ground catastrophically gave way—fire and brimstone?

As we have seen, waterlogged sand can disastrously lose all of its strength as a result of liquefaction, and it is therefore no surprise that this is the focus of intense research by engineers, physicists, and lovers of granular materials. Because liquefaction is a triumph of gravity over friction and surface tension, considerable effort has been put into eliminating the effects of gravity in order to concentrate on what's going on between the grains and the water. How? By conducting experiments on granular materials in space. By testing the behavior of sand, wet or dry, in the microgravity environment aboard a space shuttle, not only can we prepare for building on the Moon or Mars, but we can better understand the quirks of granular materials here on Earth, including the potential for better construction engineering in earthquake areas. Such experiments have been an ongoing feature of shuttle missions. Tragically, a series of such experiments was on the agenda for the crew of the ill-fated *Columbia* in 2003; some of the results had been downloaded to Earth during the mission, and readable data were recovered from the debris.

The results of orbiting and earthbound scientific investigation of granular materials contribute in a practical and potentially life-saving way to our understanding of how to handle them. Among the outcomes is better engineering to counteract, or at least mitigate, the process of liquefaction. Is one solution to try to glue the grains together so that they can't liquefy? In spite of the scale of the challenge, the answer is yes. Engineers can inject chemicals, such as epoxies, into sandy soil; however, the chemicals may be toxic. Fortunately, we may be able to seek help in this from a seemingly unlikely source. The unexpected aid comes from *Bacillus pasteurii,* a natural bacterium that lives between sand grains. Research at the University of California at Davis has shown that the bacterium causes calcite, or calcium carbonate, to precipitate, which glues the grains together. Inject sand with cultures of these bacteria, feed them well, and provide them with oxygen, and they will turn loose sand into solid rock. It's not just water and air that inhabit the spaces in between grains of sand.

AN UNDERGROUND CITY, A WORLD ON A GRAIN OF SAND

In *The Edge of the Sea,* Rachel Carson wrote:

> Walking back across the flats of that Georgia beach, I was always aware that I was treading on the thin rooftops of an underground city. Of the inhabitants themselves, little or nothing was visible. . . .
>
> In the intertidal zone, this minuscule world of the sand grains is also the world of inconceivably minute beings, which swim through the liquid film around a grain of

sand as fish would swim through the ocean covering the sphere of the earth. Among this fauna and flora of the capillary water are single-celled animals and plants, water mites, shrimplike crustacea, insects and the larvae of certain infinitely small worms— all living, dying, swimming, feeding, breathing, reproducing in a world so small that our human senses cannot grasp its scale, a world in which the micro-droplet of water separating one grain of sand from another is like a vast, dark sea.

In the great hierarchy of living things on our planet, it is generally accepted that, at the head, there are five kingdoms, one of which—animalia—contains all multicelled animals. Within each kingdom are a number of phyla: all mammals, birds, reptiles, amphibians, and fish belong to the phylum chordata. Depending on which biologist you speak to, between thirty-six and forty total phyla are recognized on our planet. Rainforests, our flagships of biodiversity, are known to contain sixteen phyla. The spaces in between sand grains are home to twenty-two.

Bacillus pasteurii, our potentially helpful construction worker, is only one of many bacteria that dwell on and in between sand grains. This way of life goes back a long way—in fact, it could go back to the very beginning of life itself. Zachary Adam, an astrobiologist at the University of Washington, has recently suggested that radioactive sand grains in beaches of the early Earth could have provided the chemical energy to assemble the building blocks of cells into the complex molecules of life. Certainly, the oldest known structures built by life on Earth were *stromatolites,* the reef-like apartment blocks of cyanobacteria (blue-green algae), which built successive layers by trapping sand grains washed across them by the primordial ocean. It was cyanobacteria that provided all-important oxygen to the Earth 3.5 billion years ago, and some forms of stromatolites survive today. As do multitudes of cyanobacteria, whose sticky filaments bind sand grains together in the soils of arid regions, forming a crust that provides nutrients and stabilizes the soil, making the task of erosion more difficult.

The dark, watery world beneath the surface of the intertidal zone of most beaches is home to many kinds of bacteria—and a jungle of other minute creatures. Silica, or quartz-based sands, are the most hospitable, since the calcium carbonate in sands composed of shell fragments makes the water too alkaline for many creatures to flourish. Pick up a handful of wet quartz sand on the beach and you are holding a miniature zoo with thousands of inhabitants. Some of these were first observed by Antony van Leeuwenhoek as he peered down his microscope at sand grains and his animalcules, but we owe our understanding of the incredible diversity and importance of this community to the work of Robert Higgins, who since the 1960s has devoted his life to identifying and describing its members. Now retired from the Smithsonian Institution, Higgins is the pioneer of work on *meiofauna* ("lesser animals"), creatures whose giants are 1 millimeter long; he continues to contribute to the identification of new species, which crop up at the rate of a dozen or so every year. In a 2001 symposium that paid tribute to his discoveries, Higgins modestly described his work as "how the 'lesser-knowns' became better-known."

Life between the sand grains is not easy. The grains move and settle under the pressure of the waves and the incessant flushing of the tides, water occasionally drains out completely, and the ecosystem contains a variety of predators. Life has had to adapt, and it has done so in strange and extraordinary ways that reflect an intimacy with the behavior of granular materials. Many of these creatures have armored or padded bodies designed to withstand abrasion by moving sand grains; their shape is often flat or elongated to enable squeezing between the grains; and they have developed a variety of ways of attaching themselves, using glue or suction, to individual sand grains, clinging on for dear life. Figure 15 is a portrait of some of Higgins's weird and wonderful creatures. Rotifers, nematodes, mystacocarids, tardigrades, gastrotrichs, turbellarians, and kinorhynchs—it's tempting to view these little animals as rejoicing in their exotic names. Tardigrades, which live in both marine water and freshwater, can't swim, so hanging on to a sand grain is vital; some use mechanical suction toes, some claws, some both. A gastrotrich can glue and unglue itself in an instant. A kinorhynch is ungainly, described by Higgins as an umbrella in a canister, but it moves effectively, if slowly, exploring one sand grain at a time. Rotifers, so named because they look like rotating hairy wheels, are represented by 2,500 different species, most of which are freshwater dwellers. Tardigrades have the remarkable ability to suspend operations if the water disappears, remaining dormant and dehydrated for a hundred years, only to spring back to life when rewetted. They seem to do this by replacing the water in their cells with sugar, which renders them immune to freezing and radiation, a talent that is of considerable interest in the worlds of medicine and extraterrestrial biology.

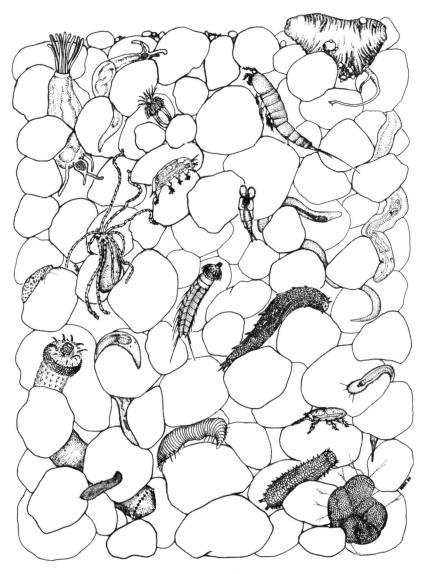

FIGURE 15. Portraits of the meiofaunal community. (Illustration courtesy of the Department of Invertebrate Zoology, National Museum of Natural History, Smithsonian Institution)

In the entire living world, only three new phyla were identified in the twentieth century, and one of them came from the strange world of meiofauna. In the 1970s, Reinhardt Kristensen, a professor at the University of Copenhagen and an old student of Higgins, showed him a collection of creatures from the coast of Brittany that he could not identify. "Oh," responded Higgins, "I have one of those too." The animals fit with no known group of living things, and it took several years of meticulous collaborative work to define the character of the creature as a new phylum. It looks rather like a classical amphora with snakes emerging from it, and they named it *loricifera,* "corset-bearing," reflecting the appearance of the rings that sheathe the animal. More than seventy species have now been identified, but we know next to nothing about their behavior, since, sadly, they die before reaching the laboratory. One remarkable thing we do know is that loricifera have the smallest cells of any known animal.

Inevitably, the community of organisms growing on or living in sand has its own name: *psammon.* Among the psammon are psammophiles (or arenophiles, if you prefer Latin to Greek), psammobionts, and psammoxenes. The names are weighty, but the members of the community are not; what they *are* is wondrous. Most of us don't know they exist, but we should be grateful for them. Without meiofauna, the sands of our beaches and lakeshores would be stinking, toxic places, with organic debris rotting unconsumed and dangerous bacteria rampant. The microscopic creatures of the meiofauna feed off this debris: they keep our beaches clean.

OUR UNKNOWN WORLD

Contrary to Heraclitus's claim, sand illustrates for us the *limitations* of our power in the world: it reminds us constantly that what we know is eclipsed by what we don't know. Whether we're concerned with the behavior of a simple pile of sand, the internal structure of a sand castle, the mathematics of criticality, liquefaction, or the daily habits of loricifera, we have barely scratched the surface of the mysteries of this strange material. And these mysteries continue to crop up in our daily lives. The pictures reproduced in Figure 16 were submitted to the *New Scientist* for its weekly section devoted to readers' questions. On a beach along the shores of Lake Michigan appeared the mysterious natural sculptures in the sand

FIGURE 16. Sand mysteries: natural sand sculptures from the shores of Lake Michigan (top) and from a Scottish sand dune (bottom). (Photos courtesy of Steven C. Thorpe and Noreen Slank, top, Martyn L. Gorman, bottom)

shown at the top. It seems that something bound the sand grains together and the loose sand around them was then eroded. But what caused the grains to bind is anyone's guess—suggestions ranged from worm excretions to raindrops, from sand boils to urinating animals. The sculpture on the bottom appeared one morning on a Scottish sand dune; what went on to produce it is again elusive. Another of the simple—and not so simple—mysteries of sand.

Sand and Imagination I

Very Large Numbers of Very Small Things

> The Walrus and the Carpenter
> Were walking close at hand;
> They wept like anything to see
> Such quantities of sand:
> "If this were only cleared away,"
> They said, "it would be grand!"
> "If seven maids with seven mops
> Swept it for half a year,
> Do you suppose," the Walrus said,
> "That they could get it clear?"
> "I doubt it," said the Carpenter,
> And shed a bitter tear.
>
> Lewis Carroll, *Through the Looking-Glass
> and What Alice Found There*

THE LOOGAROO'S OBSESSION

If you live or travel in the Caribbean and are of a superstitious disposition, or if you simply believe in the precautionary principle, take enough sand from the beach to make a good-sized pile outside your door. This will prevent the loogaroo from entering. The loogaroo is a vampire, most commonly a female. Each night, she rids herself of her skin, hides it under a tree, and flies off in search of blood, flames shooting from her armpits and orifices, leaving a luminous trail through the sky. She can take on different forms and gain entry through the slightest crack, but she has, fortunately, a weakness: she is an obsessive counter. On finding the pile of sand you have left, she will be compelled to interrupt her bloodthirsty quest and count the grains. If you have left enough, this will take her till dawn, when she must rush off to reunite with her skin. If you are concerned that there might not be enough grains of sand, mix in some nails of a ground owl—whenever she encounters one of these, she is so distressed that she drops the sand she has counted and has to start over again.

The loogaroo is clearly related to the loup-garou, the French werewolf, and is probably the result of an interbreeding of that creature and voodoo (in which a pile of sand is the home of the spirits and the symbol of the Earth). The loup-garou enjoys the bayous of Louisiana, where a sifter placed outside your door will cause it to compulsively count the holes. It would seem likely that selecting a sifter with small holes and at least trying to fill it with sand would prove doubly effective.

Compulsive counting appears to be a traditional failing of vampires in a wide range of cultures—presumably Count von Count from *Sesame Street* is not simply a play on the word. Sand is deeply associated with numbers, particularly large or extremely large ones, and counting sand is associated with the challenge of handling such numbers. This challenge can, of course, be looked at as a harmless and absorbing way of whiling away some time, hence the tradition in some eastern European cultures of putting sand in a coffin to offer the dead an opportunity for a peaceful preoccupation. It can also be looked at, not surprisingly, as a waste of time: "counting sand" is a pejorative Buddhist expression used to describe those who merely study the details of doctrine rather than putting effort into the grander scheme of things. The ability to count sand is traditionally regarded as beyond mere mortals, a divine or supernatural skill. According to Herodotus, Croesus, the fabulously wealthy king in ancient Greece, decided to arrange a performance appraisal of the available oracles. The crucial test was the ability to predict, remotely, what the king would do on a defined day. The oracle at Delphi, who won the contest, prefaced her accurate prediction by declaring her credentials: "the number of sand I know, and the measure of drops in the ocean." Although it is true that counting sand grains is occasionally the task of geologists (or, in all likelihood, geology students), it is an essentially impossible, and generally futile, activity. Imagine, for example, that you poured out merely a liter or a quart of sand as your loogaroo deterrent and you decided to check its efficacy by counting the grains yourself. At the rate of one grain per second, twenty-four hours per day, the pile would occupy you for many months (even assuming that you could *say* a number such as 9,702,337 in a second).

It is worth noting, however, that counting *in* sand is part of the origins of mathematics. The abacus evolved from a flat stone covered with sand as a renewable medium for counting and calculating, using symbols drawn into the sand or small stones placed on top. The word may derive from the Phoenician word for sand, *abak,* or the Hebrew word for dust, *avak* (or *abhaq*). An ephemeral system, its geographic and cultural origins are disputed. Equally debated is the relationship between the abacus and the origins of Arabic (or, strictly, Hindu-Arabic) numerals. The story goes that eighth-century scholars in Baghdad were inspired by the work of Indian mathematicians and began to use a new system of numerals that they called *huruf al-ghubar,* "letters of sand"—or "dust." It seems quite likely that this derives from writing on a sand table or abacus. The system was introduced into western Europe by Arabic mathematicians in the eleventh century and their symbol for zero by Fibonacci two hundred years later. Does our *0* represent the depression left in the sand when a pebble is removed?

ZILLIONS AND ZILLIONS

Even small quantities of sand rapidly start amounting to significant numbers. Draw a 1-centimeter (½ in) square on a piece of paper and spread a layer of sand one grain thick so as to fill the square. Take a magnifying glass and, testing how it feels to be a loogaroo, count the grains: there will be, typically, several hundred of them. A handful of sand, before it runs through your fingers, will contain tens or hundreds of thousands of grains, depending on their size. This was unfortunate for the Sybil, gatekeeper of the underworld in Greek mythology: having been granted her wish that she should live as many years as the grains of sand in her hand, she deeply regretted her failure to request youth as well. In Knutsford, Cheshire, the first Saturday in May is Royal May Day, when sand drawings and modest sand castles are constructed outside the houses of brides. The "Royal" in Royal May Day refers to King Knut, who is said to have emptied his shoe of sand on encountering a wedding party, wishing the happy couple as many children as there were grains of sand. If, on this May Day, he was returning from his experiment at the beach—another popular legend has him demonstrating that even the king of England is powerless to turn back the tides—the couple would have had a crowded house.

Sand as a symbol of fertility crops up around the world. The Hindu goddess Parvati, the consort of Shiva, takes on a number of divine forms, friendly and terrible, but always powerful. She is sometimes represented by five piles of sand by a river, and these are worshipped by girls. As Gauri, her maternal form, she sculpted a son from sand and commanded him to guard the entrance to her cave. Shiva, not much of a family man, was prevented from entering and so he beheaded the boy; to make up for this, he attached an elephant's head, and so created Ganesh, ironically the god of good luck. Parvati, in her various guises, is a popular subject in Indian sand-sculpting festivals.

In Cambodia and Laos and other countries of Southeast Asia, New Year ceremonies require piles of sand to be made, in varying scales of size and sophistication. The numbers of sand grains are associated with the days of a happy and prosperous life or with the number of sins for which forgiveness is being asked. In other parts of Asia, sand symbolizes rain, and rituals involve participants dressed as cows and farmers hurling sand at each other.

Although the number of grains in a small pile may seem considerable as a representation of quantities of children, sins, or days of happiness, it is, nevertheless, still a trivial number in the world of sand. It is in the realm of really, really large numbers, which stretch the imagination to and beyond its limits, that sand provides the most powerful imagery. And that imagery is long and deeply established in the consciousness of many cultures. In Shakespeare's *Richard II,* the challenge of the military task faced by the Duke of York is described as "numbering sands and drinking oceans dry." Mrs. Alving, in Ibsen's play *Ghosts,* laments that "this country is haunted by ghosts—countless as grains of sand." Something that is incredibly big is often referred to as being of "biblical proportions," and challengingly large numbers are invariably evoked in the Bible through reference to sand, particularly "the sand which is by the sea shore": the number of people that the Philistines gathered to fight with Israel, the extent of Solomon's wisdom, the quantity of the seed of Abraham and David, and the number of feathered fowls rained down on the children of Ephraim, all of these are as the sand of the sea. Indeed, "the number of the children of Israel shall be as the sand of the sea, which cannot be measured nor numbered."

The imagery of unimaginable and all-encompassing quantities occurs also in the Koran: "There is not even a single grain of sand in the obscure depths of the Earth, nor any plant, blooming or withering, that is not recorded in the Transcendent Koran." In Buddhism, the quantities of sand evoked are those transported by one of the world's great rivers, the Ganges. Traditionally, a bodhisattva goes through forty stages of practice before reaching the forty-first stage of Buddhahood. The first thirty stages require worshipping Buddhas "as many as the sand grains of five Ganges Rivers," the focus then escalating until, at the fortieth stage, the number of sand grains is equal to those of eight Ganges Rivers. Buddha thought past ages "were more in number than the grains of sand from the mouth of the Ganges."

The number of sand grains cascading across the Ganges delta, the world's largest, is a measure that successfully evokes the immeasurable, but reason requires at least an attempt to quantify such yardsticks. The fact remains that, if science and mathematics are to handle the unimaginably large and inconceivably small, then some system of calculation is necessary. Archimedes recognized this need, close to two millennia before we recognized atoms, quarks, or pulsars.

As the introduction to a manuscript known as *The Sand Reckoner,* Archimedes, the son of an astronomer, wrote the following to Gelon, king of Syracuse:

> There are some, king Gelon, who think that the number of the sand is infinite in multitude; and I mean by the sand not only that which exists about Syracuse and the rest of Sicily but also that which is found in every region whether inhabited or uninhabited. Again there are some who, without regarding it as infinite, yet think that no number has been named which is great enough to exceed its magnitude. And it is clear that they who hold this view, if they imagined a mass made up of sand in other respects as large as the mass of the Earth, including in it all the seas and the hollows of the Earth filled up to a height equal to that of the highest of the mountains, would be many times further still from recognizing that any number could be expressed which exceeded the multitude of the sand so taken.

Archimedes took up the challenge of handling very large numbers. It was the third century B.C.: our system of place-based numerals would only be introduced to Europe by Fibonacci 1,500 years later, and the existing Greek mathematical notation used a complex series of symbols that were cumbersome to work with. But

Archimedes overcame this handicap in his extraordinary development of a means of calculating very large numbers.

Using measures of finger-breadths and grains of sand, Archimedes devised a basic large number, ten thousand, which he called a *myriad,* derived from the Greek word *myrios,* "uncountable." (The Greeks had another phrase for a countless multitude: *psammakosioi,* "sand-hundred," expanded by the satirical dramatist Aristophanes to *psammakosiogargaroi,* "sand-hundred-and-plenty," to describe a character's pains versus his pleasures—of which he had precisely four.) Multiplying myriads by myriads gave very large numbers indeed, each multiplication creating a new *order.* For example, in modern terms, Archimedes' system for the third order included numbers up to 10^{24}. But he continued to pile order on order up to the 10^8 order, at which point he began *periods,* with the first order of the second period and so on—he had probably lost the good King Gelon long before this. Archimedes built this system up to a number that we would represent as $10^{80,000,000,000,000,000}$. Archimedes went on to estimate the diameter of the universe (the sphere of fixed stars) from a ratio involving the diameter of the sphere of the Earth's orbit around the Sun (yes, the ancient Greeks knew this was the way it worked) and the diameter of the Earth. His result was only the diameter of Mars's orbit, but Archimedes was moving in the right direction. From there, it was easy to calculate the number of grains of sand in the universe—ten million units of the eighth order, or 10^{63}. He told the king that, while this would appear incredible to the great majority of people, for anyone conversant with mathematics, this would be a convincing proof.

How the king might have taken all this, or applied the conclusion, is lost in the mists and the sands of time. Various modern comparisons have been made between Archimedes' extraordinary number and modern estimates of the numbers of various particles in various volumes of space, and occasionally one of these calculations will be "eerily close" to that of Archimedes. That such conclusions are essentially meaningless should not detract from the ingenuity and accomplishment of Archimedes. (Sand would, in the end, contribute to Archimedes' downfall: though he had helped defend Syracuse against the Romans with various ingenious inventions, the Romans prevailed, and one of their soldiers is said to have killed him when the mathematician berated him for disturbing his drawings in the sand.)

Centuries ahead of his time, Archimedes was the first to develop a means of representing very large numbers, but our current system of powers of ten—a million, 10^6, a billion, 10^9—is more convenient. A rough calculation of the number of "grains of sand from the mouth of the Ganges," assuming a typical annual delivery, gives 25×10^{16}, or 250 million billion, a successful evocation of an unimaginably large number. The human mind seems to have some capacity limit for grasping large numbers, after which adding zeros and giving such things names (octillions, zillions, and so on) becomes merely a game in a virtual arena. And the endgame, infinity, can never be reached merely by adding more zeros or Ganges together—however many years you spend counting the sand grains pouring into the Bay of Bengal, you will never reach infinity.

In 1980, Carl Sagan, the enthusiastic popularizer of all things astronomical, kicked off one of the most enduring, entertaining, but quantitatively pointless debates about large numbers. He declared, in his television series *Cosmos,* that "the total number of stars in the universe is larger than all the grains of sand on all the beaches of the planet Earth." The calculations are ongoing and the debate rumbles on, particularly in the ethereal realms of the internet, and there are, predictably, two schools of thought. While estimates are always increasing, the number of stars is the easier number to calculate: anywhere between 10^{20} and 10^{22}. As for the grains of sand—well, it depends. What are the assumptions in terms of grain size and, indeed, what counts as a beach? Only the areas of sand above high tide, or areas underwater as well? Depending on how you choose to do the calculation, you can derive a number that is larger or smaller than 10^{22}. And if all the sand grains of the Earth are included, not just those on beaches, then it's again a different matter. Or, invoking the biblical description of the number of the seed of Abraham, "as the stars of the heaven, and as the sand which is upon the seashore," we can call it a tie. At the end of the day, it doesn't matter: Sagan successfully evoked an extremely large number. He set up a powerful comparison between two images of vastly different scale—a comparison drawn also by the science fiction writers Brian Aldiss, in his *Galaxies Like Grains of Sand* (1959), and Samuel R. Delany, in *Stars in My Pocket Like Grains of Sand* (1984).

Sagan also commented on the finite size of the universe, asserting that if you wrote out a googolplex in full on a strip of paper, the paper would be bigger than the universe. (He demonstrated this by starting, and abandoning, the process in one of his television programs.) The terms *googol* and *googolplex* are the names for numbers substantially greater than the number of sand grains on Coney Island. In 1938, Edward Kasner, the prominent Columbia University mathematician, was talking with a group of kids, pursuing his hunch that the young have no trouble visualizing very large numbers. The children generally agreed that the number of raindrops falling on New York on a rainy day would be similar to the number of grains of sand on Coney Island, and that the numeral one followed by twenty zeros was probably a good expression of that number. Kasner suggested thinking about a larger number, one followed by a hundred zeros, and asked what it might be called. It was apparently his nine-year-old nephew Milton who came up with the name *googol*—and then *googolplex,* a one followed by, according to Milton, as many zeros as you could write before your hand became too tired. A googol (10^{100}) and a googolplex (10^{googol}) were formalized by Kasner and James Newman in their 1940 book, *Mathematics and the Imagination.* While an enjoyable term, the googol is so large a number that it is not represented by anything in the known universe, whether it be sand grains, stars, or atoms; it is not widely used, but it is reputed to have been the origin for the name of a certain internet search engine.

GEOLOGIC TIME AND ETERNITY

Unimaginably large numbers of particular importance include measures of geologic time. The Earth is 4.6 billion years old; the dinosaurs died out 70 million years ago. By the human measure of three score years and ten, these descriptions of time are impossible to grasp. The term *deep time* is sometimes used to refer to the geologist's clock, as if these expanses are domains to which only geologists, through some arcane perspective, are privy. The fact is that these measures of time are as difficult for a geologist to come to terms with as for anyone else, but, of necessity, we do. Perhaps the trick is looking closely at natural processes, how slowly they work, yet what dramatic changes they bring about. Entire mountain ranges have risen, yet are destined to be worn away until only their roots remain. Entire oceans have been created and destroyed, great thicknesses of sediment have been built up from the steady work of rivers. Looking at the rates at which these processes are happening today, and assuming that they have always happened at roughly the same rate, the *only* conclusion is that the development of our planet as we see it today took a vast, almost unimaginable length of time. But imagine it we must—and we shall return to this in chapter 7.

Any number of ingenious ways of visualizing and illustrating the scale of geologic time have been devised. Condense the Earth's life span into twenty-four hours or a year and note how utterly trivial is our time on Earth: if the Earth is a year old, we appeared at seventeen minutes to midnight on December 31. If one sheet of paper represents one year of the Earth's history, the stack would be 450 kilometers (280 mi) high. A particularly vivid method, if you have a long enough corridor, is to roll out toilet paper. However, if every sheet were to represent a century, the corridor would have to stretch from London to New York and the history of *Homo sapiens* would cover the length of a Manhattan city block.

But since this book is about sand, and you are now intimately familiar with both heaps and individual grains of the stuff, perhaps a sand pile would be an appropriate illustration of geologic time. You can illustrate this yourself with a good quantity of medium-grained sand (I have simply used builder's sand). Put a single grain on the floor—this grain represents one hundred years of the Earth's history. A century is a length of time we can grasp, an aspirational but realistic life span. A cup of sand contains around three million grains, so now add three heaping cups of sand to make a pile (Figure 17, top). The Earth, as represented by our pile of sand, is now one billion years old, and, finally, the first primitive single-celled life forms emerge. Add another eleven cups of sand, and the Earth has reached the time when simple animals first occupied the land, just over 400 million years ago (Figure 17, middle). A final addition of one and a third cups brings the pile to the present day (Figure 17, bottom)—you have to look very carefully to see the difference. Remove a good pinch and the pile represents the first appearance of modern humans; take away ten teaspoons to show when the dinosaurs were wiped out.

Very large numbers present the same problem for us as a landscape without trees—we need some yardstick, some analogue to grasp to enable us to appreciate scale. We have seen in chapter 1 how our imagination can use an individual grain of sand as a measure of something minuscule, often to help us comprehend something vast. An accumulation of sand grains can similarly provide an image that resonates with large numbers, immense lengths of time—and a means of approaching the concepts of infinity and eternity.

FIGURE 17. A sandpile history of the Earth. (Photos by author)

A googolplex is no closer to infinity than it is to the number one. In 1714, Jonathan Swift, the Irish satirist, churchman, and author of *Gulliver's Travels,* wrote to the *Spectator:*

> Supposing the whole body of the earth were a great ball or mass of the finest sand, and that a single grain or particle of sand should be annihilated every thousand years. Supposing then that you had it in your choice to be happy all the while this prodigious mass of sand was consuming by this slow method, until there was not a grain of it left, on condition you would be miserable ever after; or supposing that you might be happy ever after, on condition you would be miserable until the whole mass of sand were thus annihilated at the rate of one sand in a thousand years: which of these two cases would you make your choice?

A difficult question that prompts a vague idea of eternity. The same approach was later taken by another Irishman, James Joyce, in *A Portrait of the Artist as a Young Man,* contemplating the eternity of hell—"Eternity! What mind of man can understand it?" To assist in the task, Joyce has his sermonizing priest use as his yardstick a child's small handful of tiny grains of sand. He then asks that we imagine "a mountain of that sand, a million miles high, reaching from the earth to the farthest heavens, and a million miles broad, extending to remotest space." Multiply this mountain by the number of drops of water in the ocean, and then imagine that, every million years, a bird arrives and carries off a single grain of sand. By the time the bird had removed all the sand, "not even one instant of eternity could be said to have ended." An effective calibration. And read Jorge Luis Borges's "The Book of Sand" (1975), the haunting story of a terrible, infinite book, so called "because neither the book nor the sand has any beginning or end."

For millennia, sand has had a place in our imagination to represent numbers too large to grasp, the passage of time, and eternity; at the same time, it is a symbol of things almost too small to see, the microscopic. The power of this imagery derives in part from the familiarity of sand, its ubiquity as an everyday material. It is one of our planet's most fundamental materials, shaping the surface of the Earth and our environment. The following chapters will look at those diverse habitats and the dynamics of the sands that sculpt them.

4

Societies on the Move

A Journey to the Sea

The river knows its way to the sea;
Without a pilot it runs and falls,
Blessing all lands with its charity.

Ralph Waldo Emerson, "Woodnotes II"

ORIGINS

In the time that it took you to read the epigraph above, trillions of sand grains around the world moved, taking another step on their endless journey. The Earth's business is change, ceaseless movement and recycling, a kinetic system where everything is in constant motion, be it continents plodding along at the rate a fingernail grows, or landslides and earthquakes changing our world in an instant. Sand is a principal actor in this dynamic and never-ending performance, by virtue of its size born to be transported by gravity, rivers, waves, ocean currents, and winds, shape-shifting into ever-changing landscapes. The science of geology is entirely about change. Changes with time, climate, latitude, altitude, temperature, pressure, velocity, depth, size—these are the voices in the stories the geologist seeks to read and to tell.

When I reflect on the things that lured me to geology as a child, apart from family holidays on the crumbling and fossil-strewn coasts of England, two books are clearly important. The first, out of print and not generally regarded as a classic, is *The Earth's Crust: A New Approach to Physical Geography and Geology* by L. Dudley Stamp. Published in 1951, it was called "a new approach" because Stamp had commissioned huge models of different landscapes—and what lay beneath them. The book included photographs of the models, which had been constructed in exquisite detail by Tom Bayley, a lecturer in sculpture.

For me, the book was magic. I was captivated by the models. I studied them intently and even tried, messily, to make some myself. They opened my eyes to how landscapes and their foundations work. I could look around me on those summer holidays with a different understanding—I could *visualize* things. And, best of all, some of the models depicted *change*—for example, the appearance of a glaciated valley when it was still filled with ice and after it had melted (illuminating for family excursions in the ice-sculpted landscapes of Wales and Scotland). My curiosity about the world around me was aroused; I was alerted to geology, to the processes of the Earth's surface.

It is these processes, constantly performed by rivers and oceans and wind, and involving mass migrations of societies of sand grains, that we will look at in the next three chapters. But first, the second book. This *is* a classic, and I was its proud owner, at the age of six, in Minneapolis. It was only as an adult that I realized its importance to me in terms of revealing the world and of setting the scene for the journeys of geology. It was *Paddle-to-the-Sea,* and it will resonate particularly with North American readers of a certain age. Holling Clancy Holling was a Michigan-born writer and illustrator of children's books, by far the most acclaimed and loved being the story of the journey of an Indian boy's small wooden carving of a canoe and its occupant. *Paddle-to-the-Sea* was first published in 1941 and reprinted many times.

In the hills above the Canadian shores of Lake Superior, the boy carves the canoe to take the journeys that he cannot, and on the bottom he inscribes the words "Please put me back in water—I am Paddle to the Sea." Setting the carving on a snowbank in spring, he watches the meltwaters carry Paddle down into the streams that flow into Lake Superior, the start of a meandering four-year journey to the sea. Paddle's adventures through the Great Lakes to the Gulf of St. Lawrence and the Grand Banks are not only a gripping narrative, exquisitely illustrated, but later the reader realizes how much he or she *learned* along the way. And Paddle's journey, as the canoe is swept around by currents and waves, marooned for long periods on beaches and sandbanks—and by human activities—before traveling on, is an allegory for the movement of sand.

Whenever our family car would break down en route to our vacation destination, my father would cheerfully declare that "it is better to travel hopefully than to arrive." While as a child I was never fully convinced of this, the statement (paraphrasing Robert Louis Stevenson) does evoke the nature of journeys, human and granular. In this chapter, we will take a river journey with a sand grain, a journey driven primarily by natural forces but also one that, for the great majority of the world's rivers, inevitably encounters human influence. Rivers are highways for sand and humans, and interaction is as unavoidable for the sand grain as it was for Paddle.

BEGINNING THE JOURNEY

Pick up a single sand grain at the beach and you have encountered it in the midst of a potentially endless journey over vast distances and immense lengths of time. It has been swept around the surface of the Earth by rivers, waves, and the wind. It has been buried, compacted, soldered together with its companions into solid sandstone, melted, uplifted, battered, and swept away again, pausing, perhaps, to be excavated, mined, or quarried. The sand grain's tale is like that of a soldier—long periods of boredom punctuated by brief episodes of brutal activity. Except that the long periods are truly long, and the episodes are often on an immense scale.

In chapter 1, we saw the birth of an individual sand grain, liberated from solid granite by the relentless attack of weathering. Falling onto the slope beneath its parent rock, nudged or swept downhill by rain and torrents, the sand grain joins countless traveling companions under gravity's rule. Most sand grains on Earth have taken a journey down a river at least once in their lives, and perhaps hundreds or thousands of times.

Think, for example, of a sand grain born, for at least the second time in its life, in one of the most remote and spectacular places on Earth, the Canaima National Park in southeastern Venezuela, near the borders of Guyana and Brazil. It has been resting for some hundreds of millions of years in a sandstone that is one of the oldest in the world. Two billion years ago, great rivers flowed across this part of what we now call South America, meandering, flooding, constantly changing their routes, just as the Amazon and the Orinoco do today. And, as rivers always have, they carried with them huge cargoes of sand, traveling hopefully, born from the disintegrating rocks of ancient mountains long since subdued and flattened.

These ancient South American rivers flowed, as most rivers do, over the continent to an ocean. But this early version of South America looked very different than it does today; small ancient fragments of continent had only recently been welded together to form the kernel of what would become today's continent. Close to a couple of billion years would pass before the Andes were plastered along its western edge. The ocean to which the rivers flowed has long since been swallowed up and obliterated by the voracious churning of the Earth's plates, today's Atlantic being only a distant, indirect descendant.

As the rivers slowed on approaching the sea, as floods abated each year and channels changed their course, the rivers lost the energy to carry their cargo of sand and dumped it in sandbars, estuaries, deltas, beaches, and offshore banks. This process continued, year after year, over millions of years, and these abandoned layers of sand eventually covered a vast area of the fledgling continent. As the rivers roamed over the landscape, the layers built up in places to an enormous thickness, their foundation sinking to accommodate the load. Buried under this great pile, the sands were compacted and solidified into hard, rocky sandstone.

Much, much later, the continent, buffeted by colliding plates along its edges, was heaved upward again. The sandstones were back at the Earth's surface and began their battle with the elements, a campaign that they continue to wage today, hundreds of millions of years later; these two-billion-year-old sandstones form the extraordinary landscapes of the Canaima National Park. Vertical cliffs a kilometer or more high protect vast flat-topped plateaus that look out over hundreds of kilometers of forest and savannah. The layers of sandstone once covered these areas, but they lost the battle with the elements and were stripped away. The mountains are the resistant remnants and have probably looked very much the way they do today for more than three million years (Figure 18).

Incredibly, in spite of all the geological violence going on around them—the split from Africa, the opening of the Atlantic, the rise of the Andes, and the splintering of the Caribbean—these sandstone layers remain essentially as they were deposited two billion years ago, not tilted but horizontal. It is this that creates the great flat-topped peaks, dissected and isolated by erosion around them, but toweringly intact. Sir Arthur Conan Doyle set his story *The Lost World* on these virtually inaccessible plateaus (*tepuis,* table mountains, in the local Pemon language), imagining them with dinosaurs and sandbanks "spotted with uncouth crawling

FIGURE 18. The table mountains of the Canaima National Park. (Photo from Stewart McPherson, *Lost Worlds,* McDonald and Woodward Publishing Company)

forms, huge turtles, strange saurians, and one great flat creature like a writhing, palpitating mat of black greasy leather, which flopped its way slowly to the lake." The true ecology of these plateaus is only somewhat less incredible. Long isolated from larger communities, many of the amphibians and reptiles that live here are found nowhere else in the world, including blind black frogs a couple of centimeters long. Nine hundred species of higher plants have been identified on one mountain alone, of which 10 percent are endemic. Thirty species of birds are unique to the area, and the task of describing the character and diversity of the insect life will continue to occupy entomologists for a long time. Many of the mist-shrouded summits in the park, which was awarded World Heritage status in 1994, remain completely unexplored.

So, a grain of sand, torn from its mother rock and mountain roots more than two billion years ago, carried along by ancient rivers, dumped in a sandbar, buried under a great thickness of more sand, heated and squeezed together with its companions, and elevated a couple of thousand meters above sea level, has resided there ever since. The climate on the plateaus of the tepuis is extreme: an average of 2.5 meters (8 ft) of rain fall each year and temperatures range from freezing to 30°C (90°F). The torrential volumes of rainwater form pools, bogs, and streams, and the plants, many of them carnivorous, combine with these to concoct a mildly acid brew. The constant warming and cooling, the chemical effects of the waters, the root systems of the plants, and the general extremes of a tropical climate conspire to attack the integrity of even the oldest and hardest sandstones. As a result of this incessant weathering, the rocks are sculpted and worn into monoliths, towers, sinkholes, gullies, and canyons, within which water constantly pools and overflows.

The sand grain, which last saw individual freedom two billion years ago, is finally liberated by the incessant onslaught of climate and chemistry and carried by the rain to a small pool. During a particularly heavy downpour, the pool fills and overflows, and the sand grain is washed out. Tumbling down a gully in the torrent, it is unceremoniously swept over the edge of the plateau in a plummeting waterfall; the drop is a kilometer (3,000 ft) or more over Angel Falls. The falls are named not in recognition of their apparent origin in the mists of the firmament, but after Jimmie Angel, a flamboyant bush pilot seeking gold in the footsteps of the conquistadors and Sir Walter Raleigh. This region of Venezuela is one of the reported locations for the mythical city of El Dorado, and there certainly is gold there, but Jimmie Angel never found it. On one of his trips in 1933 he did, however, see a huge waterfall, cascading "from the sky"; flying to its top and base, Jimmie accurately estimated its height. Angel Falls is the highest free-falling waterfall in the world, and much of the water evaporates by the time it reaches the bottom. But it carries our sand grain—together with its companions—on its new journey of freedom.

The grain of sand, having hurtled off the plateau, finds itself in one of the streams feeding into the Rio Caroni. The Caroni has risen farther to the south in the Canaima and flows northward, joined later by the Paragua, and onward toward the Orinoco and so out to the Atlantic. All the water raining on an area of mountains, forest, and savannah larger than Scotland finds its way into the Caroni, and the flow of the river in flood is as much as 10,000 cubic meters (350,000 cu ft) per second—

close to the average flow of the Mississippi at New Orleans. In a massive movement of water like this, even when the river is not in flood, the sand grain has virtually no chance to rest. The turbulent water picks it up, carries it along, drops it, and picks it up again, swirling and hurtling onward. The total volume of sand and mud transported in this way by the river is huge: 52 million tons every year, on average. To put it another way, if you were to stand by the river for a day, 5,700 large dump trucks full of sand and mud would drive past. And this isn't even the Orinoco, never mind the Amazon.

The Caroni would seem to be a perfect example of a wild river, a case study for the natural processes of a river transporting sediment. But, sadly, it is not. Our sand grain's journey, which began so well, so dramatically, is doomed.

Within 150 kilometers (100 mi) of the start of its journey, instead of continuing its tumultuous travels down the riverbed, the sand is dumped, suddenly and unnaturally, into an enormous lake. This is Lake Guri, at an area of 4,300 square kilometers (1,700 sq mi) one of the world's ten largest reservoirs. Fifty years ago, it didn't exist. The lake is formed by the Guri Dam, the second (soon to be the third) largest hydroelectric generating plant on the planet, and it is the end of the road for our sand grain. The grain will remain trapped in the lake behind the dam, buried along with millions of tons of its fellow travelers. The massive water flow in the Caroni is causing the river's demise: there are three further dams on the river below Lake Guri, and three more are planned above it.

But the consequences of stopping a sand grain on its journey are huge. The effects of dams, not only in the Orinoco system, but anywhere in the world, include major modifications to the delivery of sediment to the river system downstream and to the ocean; this, in turn, alters entire landscapes. Changing with the seasons, a river system left to its own devices irrigates and drains, excavates and backfills the terrain through which it flows. Disturb that delicate but massive balance by disrupting water flow and impounding sediment, and you change landscapes and ecosystems. Louisiana loses over 30 square kilometers (12 sq mi) of critical wetlands every year as a result of the reduction in sediment carried by the Mississippi and the effects of levees and canals. This is the land that should not only maintain biodiversity but also, critically, act as a natural defense against hurricanes. Such changes, however, will be dwarfed by those stemming from the Three Gorges project on the

Yangtze River above Shanghai. The reservoir there will be 600 kilometers (370 mi) long when it is full and a very effective sediment trap. The amount of sediment coming downstream has more than halved since the dam's completion, and the delicate balance between sediment deposition and removal at the mouth of the river has been destroyed. The city of Shanghai, built on sandbars that emerged only in the thirteenth century, depends on this balance to maintain its defenses against the sea. In the United States, large dams have been built at a rate equivalent to one a day since the signing of the Declaration of Independence. Today, increasing effort is being put into removing them.

ARTERIES

The world's rivers are the arteries of sand transport across the continents. Rivers, *fluvial systems,* are immensely complex and varied and responsible for delivering huge quantities of material to the coasts and all other sedimentary systems. They occur everywhere, in every climatic zone, on every scale, and in every kind of terrain—and this has been true throughout the Earth's history since the first rains fell. Among their hydraulic complexities is the way in which they transport sediment—hence Albert Einstein's warning to his son (see chapter 1). Ralph Bagnold's pioneering work, after he had moved on from wind transport problems, took matters back to first principles; he defined, in the language of physics, the *power* of a river, and observed that it can be viewed as an engine, doing work, transferring energy from the fluid to the grain. And the world's rivers are, indeed, magnificent engines.

Although human interference through dams and other activities is increasingly disruptive, if the Earth were left alone, on the order of 8 cubic kilometers (10.5 billion cu yd) of the material of the continents would be swept away by rivers into the oceans *every year*—that's half a billion dump truck loads. By some estimates, around a third of this natural volume is prevented from reaching the oceans as a result of being trapped behind dams and other man-made obstructions. Nevertheless, the continents are still being constantly ground down, dissolved, cleaned off. Every thousand years, the Himalayas are reduced in height by 30 centimeters (12 in), and the area drained by the Mississippi is lowered by 10 centimeters (4 in). These rates may seem small, but the process never stops. James Hutton, one of the founders of geology, declared in his *Theory of the Earth* (1785): "The surface of the land is made by nature to decay. . . . Our fertile plains are formed from the ruins of the mountains."

The Caroni is one of the world's large rivers—but it is dwarfed by, among others, the Yellow and Yangzte, the Ganges and the Brahmaputra. The size of a river is a matter of what you measure—length, drainage area, sediment transport. It is that last yardstick that we are interested in here. Many factors of terrain and climate contribute to this measure, and it is the most difficult to calculate. For the engine operates in three different ways to carry its sediment load: by rolling and bouncing grains, pebbles, and boulders along its bed, by swirling smaller grains into almost continual suspension, and by carrying soluble minerals dissolved in the water. The last, the *dissolved load,* is simple to measure, the *suspended load* relatively simple, and the *bed load* extremely difficult. Measures of the sediment load of rivers commonly are for the suspended load only, but the bed load is huge. And all of these numbers are only averages—a river in flood will carry orders of magnitude more sediment than it does on a normal day.

To understand some of these processes, we can sometimes turn a river into a laboratory for experiments on a grand scale. In 1963, the Glen Canyon Dam was built 150 kilometers (100 mi) upstream from the Grand Canyon. It did what dams do well and trapped more than 90 percent of the sediment carried down the river, with the result that the canyon was starved of sand. The clear water below the dam had a significantly increased capacity for sediment transport and swept away old sandbanks, reducing habitat for flora and fauna, not to mention tourists. Species of fish that had relied on the protection of murky, sediment-filled water vanished, and others were left vulnerable. Experiments were designed and carried out to release vast volumes of water and sediment from the dam, which would flush out further sand and silt that had become marooned in side canyons, rebuilding the natural sandbanks. The first experiment, in 1996, was overeager—the flow of water was so great that all the sediment was flushed, unimpeded, 450 kilometers (280 mi) down the canyon to Lake Mead.

The second, more controlled experiment, in 2004, was monitored in every way imaginable by the experts. Some scientists traveled by boat along with the peak flow, measuring how the sediment content of the water changed as it surged downriver. One researcher from the U.S. Geological Survey described it as "like riding a bus and keeping track of how many people are getting on and off, but we keep track of how many sand grains stay in that packet of water" (O'Driscoll, 2004). While this experiment was more successful than the first, its benefits were limited, in part because only modest amounts of sediment were available. By March 2008, the volume of accumulated sediment was three times that in 2004 and the experiment was repeated. The taps were opened by the interior secretary, who declared, "the water will be released at a rate that would fill the Empire State Building within twenty minutes," and "it will transport enough sediment to cover a football field one hundred feet deep with silt and sand" (Environmental News Service, 2008). Analyzing the results will take years, and the durability of any positive results remains uncertain.

Not only floods, but also other unusual natural events dramatically change a river's burden. Earthquakes frequently trigger landslides, vastly increasing the amount of sediment cascading into, and carried away by, the area's rivers. The eruption of Mount St. Helens quadrupled the amount of sediment carried by the Columbia River, and the river's navigation channel was reduced to a third of its pre-eruption depth. In the years following the eruption, nearly 110 million cubic meters (140 million cu yd) of sediment were removed from the main river channels by the U.S. Army Corps of Engineers.

Rivers are naturally complex and highly variable. Our imprint simply adds to the complexity in, as we have seen, a far from modest way. To follow a sand grain on an unimpeded natural journey is not easy—even if the journey begins, like Paddle's, in a wilderness area. As a consequence of the Aswan Dam, the Nile Delta now receives virtually no sediment at all, and Mediterranean waves and currents are reducing its size daily. Close to 80 percent of the water discharge of the largest river systems in the Northern Hemisphere is affected by human interference. Unaffected rivers lie almost exclusively in the Arctic. Along with water discharge goes sediment transport, and the consequences of our interference, planned or not, are often dire. But it's not all bad news for sand: reservoirs fill up, and the natural flow of sediment, after a long time, resumes. Major floods can scour reservoir sediment and flush it through the dam system. The Mississippi still ranks in the world's top ten of suspended-sediment deliverers, in spite of the dams and other human interference along its course.

So, where shall we go to follow the *almost* unimpeded river journeys of societies of sand grains? To a river that winds its way through dramatic landscapes but is far from wild, a river via which a sand grain has a good chance of reaching the sea to continue its travels. To the "long winding river," the "long reach river," or the "great island river," depending on which Native American language you choose: the Susquehanna. A sand grain's journey down the river follows the path of Jack Brubaker's book *Down the Susquehanna to the Chesapeake,* a compelling biography of a river.

JOURNEYS DOWN THE LONG, WINDING RIVER

Twenty thousand years ago, the northeastern United States was covered in ice. The temperature had plunged during the last of the series of glacial periods, and massive ice sheets had invaded, grinding their way unopposed from the north. They had sucked up volumes of the planet's water, and sea level was over 100 meters (300 ft) lower than it is today. Setting out from where Atlantic City would eventually be, you would have had over 150 kilometers (100 mi) to walk before you reached the shore. And the ice, grinding and scouring, had brought with it an unimaginably huge cargo of debris, gouged out from the lands to the north. The ice sheets moved southwestward, creating havoc as they went. The Susquehanna River had been carving out its sinuous valley for a long time before this and previous incursions of ice. Where the valley paralleled the direction of ice movement, it was scoured out, but in sections where it was perpendicular, the glaciers simply overrode the intervening hills and filled the valley with debris. Whole regions were dammed by debris, and ice and lakes formed behind the dams.

When the climate warmed and the ice began to retreat, around nineteen thousand years ago, the situation only got worse. As if pursuing a strategy of leaving chaos in the wake of their retreat, the ice sheets dumped every boulder, clay particle, and sand grain that had been trapped within them, disfiguring the landscape. As the ice withdrew, the waters dammed in lakes were liberated, and huge floods hurtled down the valleys. The rate at which the ice retreated, nearly 2 kilometers (over a mile) roughly every fifty years, was so rapid that the land had no chance to adjust. Up until thirteen thousand years ago, an ice dam in the upper reaches of the Hudson River valley had held back a lake three times the size of today's Lake Ontario; but the dam collapsed, and the lake level dropped rapidly by 120 meters (400 ft). The flood was catastrophic, carrying car-sized boulders past Manhattan and beyond the (then distant) shoreline. The sudden huge input of freshwater into the Atlantic was sufficient to change the climate. Out of this chaos farther south emerged the Susquehanna River of today, patiently reestablishing itself, carving through the glacial debris in its valleys, conspiring with each one of its tributaries to sculpt today's landscape.

When rain falls in central New York State and nudges a sand grain downslope, the grain's future journey is determined by the landscapes that the ice sheets left behind. The sand grain could be starting a journey northward to the Mohawk River, then on to the Hudson, and Manhattan; northwestward to Lake Ontario, past Montreal, and out to the Atlantic via the St. Lawrence; or southward into the Susquehanna and a long, leisurely 700-kilometer (440 mi) trip to the Chesapeake Bay and, perhaps, the ocean. Which journey is taken depends on exactly where the raindrop falls, which valleys still remain blocked by glacial rubbish, and which have been opened. Otsego Lake, officially the source of the Susquehanna and a source of wonder to Natty Bumppo in James Fenimore Cooper's *The Deerslayer,* breached the glacial dam at its southern end and drains southward, but the great Finger Lakes, not far to the west, remain blocked at their southern ends and drain northward into Lake Ontario. The area within which all water—and all sediment—drains into the Susquehanna covers 70,000 square kilometers (27,500 sq mi) of New York, Pennsylvania (nearly half of the state), and Maryland: this is the largest drainage area of any river on the East Coast (Plate 6).

Our Susquehanna sand grain, plucked by ice from an ancient Canadian terrain and liberated from the glacial sediment in which it had been entombed for ten thousand years, is carried into Otsego Lake, flushed out by a major flood, and carried out into the upper reaches of the Susquehanna. It exits the lake at Cooperstown, New York, home of Fenimore Cooper and baseball.

The river is still today adjusting from the devastation of the ice sheets, eroding away material here, depositing it there; the land itself is still recovering from the burden of ice, rising in some areas, sinking in others, and these are further changes that the river must respond to. The river falls 365 meters (1,200 ft) on its journey from Otsego Lake to the Chesapeake, and if all were stable, its *profile,* the shape of the curve plotting the topographic elevation of the river bed versus distance from the source, would be smooth and concave upward, its steepness gradually decreasing downstream, toward sea level. The river would be in equilibrium, delivering exactly the same amount of water and sediment that it receives, erosion and deposition in perfect balance. But it is not an ideal world, and the constant adjustments that the river has to make mean that it will be a very long time—if ever—before it achieves stability. In its constant prehistoric battles with ice, sea level, and the ups and downs of the land, the Susquehanna has developed a distinctly peculiar profile. Add to the natural processes man-made changes that liberate more sediment—deforestation, agriculture, and mining, for example—and it is clear that the Susquehanna is a very dynamic river, constantly shifting its course, eroding its banks, and dumping its sediment. This dynamism is complex in the extreme, but there are fundamental principles that we can observe, measure, and replicate.

RIVERS AT WORK

The ways in which rivers operate—or, to use the engine metaphor, literally *work*—cover a huge range of scales, involving the smallest grain of sediment to the topography of a continent, what happens in a second to changes over millions of years. To begin at the small scale, the world of our sand grain and its companions, we can ask the apparently simple question, "How can we predict whether that grain will be transported or deposited?" The answer is not so simple, but we can define the principles. The sand grain in river water finds itself at the mercy of several different forces. Always there is gravity, urging the grain to fall. Then there is the frictional drag force of the flowing water, urging the grain to accompany it on its downhill journey, but at the same time slowing it down. The grain may be rolled along the bed of the river, or it may be lifted by the same force that lifts an airplane; once plucked from the riverbed, the grain will be carried along by the swirls and eddies of the turbulently flowing water, which accomplishes its work by transferring its energy to the grain.

How far the grain will be carried once it is lifted into the flow depends on a number of factors: grain size, weight, and shape, as well as flow speed. This is where Bagnold's work—again—comes into play. His definition of particle size according to the rate at which it settles in water is all-important in determining how far a particular grain may be carried. In still water, a coarse sand grain will fall about 10 centimeters (4 in) in a second, but a clay or mud particle will take hundreds of years to fall the same distance. In a river, both will be carried sideways as they fall, moving downstream until gravity causes them to settle back to the riverbed—where the sand will hit other grains and nudge them into movement. The grains move in leaps and bounds, a process called *saltation* (from the Latin verb *saltare,* "to jump" or "to dance"). If the grain is small enough and the current fast enough, it will be suspended, along with its muddy colleagues, and travel long distances, potentially for its entire journey.

A river of a given volume of water flowing at a given speed has the capacity to transfer its energy to masses of moving sediment. The capacity to do so increases exponentially with the speed of the river's flow, in proportion to the *cube* of the amount by which its velocity exceeds that necessary to start sediment movement. Thus small increases in the velocity can cause the mobilization and transport of large incremental volumes of sediment. But the capacity is not limitless. There is always a constraint to the transporting capacity, and if the capacity drops—if the flow slows—then so do many of the grains, and they are deposited on the bed of the river. The physics of these processes acting on one grain can be worked out, but the scaling up to the flow of billions of grains is beyond computing power—particularly when the interactions between each and every one of them must also be taken into account. Instead, we resort to playing in the sand. We build laboratory-sized rivers, water tanks, and sluices, vary the different factors, and watch what happens—just as we can on a beach.

There is one research facility that takes this idea of playing with sand very seriously indeed. At the St. Anthony Falls Laboratory, on the banks of the Mississippi in Minneapolis, large volumes of water are drawn from the river to be fed into "one of the most expensive sandboxes in the world" (National Center for Earth-Surface Dynamics). This experimental apparatus works on an extraordinary scale: appropriately dubbed "Jurassic tank," it is half the size of a tennis court and holds two hundred tons of sand. Not only can the overall slope of the tank be varied, but the shape of its floor can be molded into a small-scale version of any natural topography over which water and sediment might flow. The original Jurassic tank has spawned a variety of other specialized experimental apparatuses: the main river model basin is 47 meters (155 ft) long and the depth of a bathtub. Out of these extraordinary constructions have come extraordinary results. Erosion, transport, and deposition of sand can be observed as they happen, and the resulting deposits can be dissected and analyzed in any number of ways to contribute to our understanding of natural-scale systems.

The amount of sediment transported by a river is its load; if the river has the capacity, it will add to the load, scouring and eroding its banks and channel, whereas if its capacity drops, it must dump some of its load. A river's capacity will vary with shifts in velocity as it follows its own bends or encounters changes in the shape of its bed. As when a cup of coffee is stirred, when river water flows around a bend, it is flung to the outside, its velocity increases, and it carves away at its bank. On the inside of the bend, the velocity and the load-carrying capacity drop and sediment is deposited. The sandbanks that build up on the inside of river bends are termed *point bars*. They may be swept partially away as capacity rises in a flood, but as the flow subsides they build up again—point bars are long-term features. Each episode of sandbar construction is marked by a new set of layers, with larger grains at the bottom of each layer and smaller grains at the top, reflecting slowing current velocity; the sand is graded (Figure 2).

The sands in point bars tell us important stories, not only of modern rivers, but also of ancient ones. As a point bar grows out into the river, it also encroaches downstream: sand is transported across the submerged part of the bar and tumbles down the downstream slope, which is built out with each flush of sand. As a result of this, the layering of the sand within the bar takes on the angle of that slope; while the bar as a whole may be flat, its internal structure consists of layers that gently slope downstream. The layering of a sediment or sedimentary rock is termed *bedding,* and it tells us a great deal about the way in which the sediments were deposited. The large-scale bedding in a point bar is flat, but those layers are constructed from individual small-scale sloping beds of sand—*cross-bedding* (Figure 19). There are many subtly different kinds of cross-bedding, but all result from

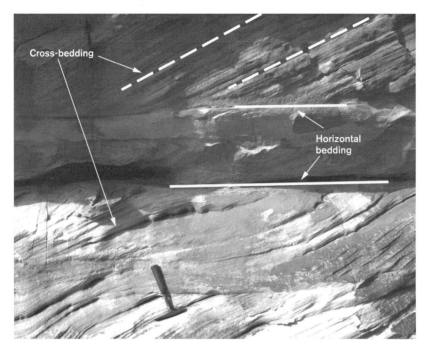

FIGURE 19. Cross-bedding in ancient sandstones. The main bedding is horizontal, but internally the bedding is at an angle, reflecting the underwater slope of the migrating front of a sandbar, moving with the current from right to left. (Note the hammer for scale.) (Photo by author)

sediment deposited by a current and give the geologist key information about the direction and strength of the river's flow—today or hundreds of millions of years ago (chapter 7).

GREAT BENDS

No river is perfectly straight for any significant distance, and bends, sometimes wildly sinuous and exaggerated, are the norm. In favoring a winding route, the river is said to *meander*. This is a highly dynamic process. As it constantly wears away the outside of the bend and deposits sand on the inside, the river channel migrates sideways (Figures 20, 21).

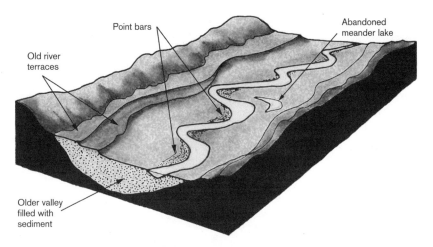

FIGURE 20. Meanders, point bars, and river terraces in a river valley.

FIGURE 21. A river, meandering with point bars. (Photo by author)

As the meanders migrate and exaggerate, they may even cannibalize one another and create an entirely new course for the river, abandoning the old one. Meandering rivers are often accompanied by curving patterns in the surrounding land, remnants of the old channels and point bars, palimpsests of the landscape. The only lateral limits to the migrating channel are the sides of the valley; a typical meandering river will "ricochet" from one side of the valley to the other over thousands of years. A segment of a channel may change its course overnight, shifting its position by hundreds of meters a year. Given that large numbers of people around the world live in the shifting settings of river valleys and that significant economies depend on them, understanding these dynamics is vital. Prior to major human interventions, meanders of the Mississippi were migrating laterally at up to 60 meters (200 ft) every year. Tributaries of the Ganges have shifted 150 kilometers (over 90 mi) in a couple of centuries, and flooded residents of Bangladesh recount how the nearby overflowing river channel was far away when they were children. Shifting sands constantly rearrange the planet—and the lives of its inhabitants.

As our sand grain and its companions roll and skip down the Susquehanna bed or leap up into the river's flow, they make a brief preliminary foray into northern Pennsylvania and approach the town of Great Bend (Plate 6). The river winds its way through the Endless Mountains of the Allegheny Plateau, rounded hills whose tops mark ancient bedrock protruding from the blanket of glacial debris. The valleys have been deeply cut by the river and its tributaries as they reestablished themselves after being rudely disrupted by the ice. (It is, incidentally, along this stretch of the river that Joseph Smith translated most of the Book of Mormon from the gold plates that were reportedly found buried in the glacial sediments of north-central New York.)

The river is confined to a relatively narrow valley here, but its meanders swing from side to side, eroding glacial sediments and bedrock on the outside of the bends, depositing broad point bars on the inside. Here we see human dependency on the dynamics of river sand: where the Susquehanna takes its great bend northward again, back into New York, the flat land of the resulting point bar has been exploited by both the town of Great Bend and the railroad.

If, as the Susquehanna meanders its way back into New York before finally committing itself to its southward journey through Pennsylvania, you were to stop on any point bar and dig a trench, in the walls you would see the sloping layers of cross-bedded sand. In many places, if you were to rest from your labors and look up, across the river, to where walls of bedrock have been exposed by erosion, you might well see the same structures in the solid rock. If this were to happen, you might be looking at another point bar, one that was constructed in a river that flowed through this area 370 million years ago, when huge river systems flowed westward from the rising terrains of continental collisions to the east (see chapter 7). Our sand grain skips and dances past its ancient relatives, perhaps to be joined by some of them, liberated from the rock by the flowing river.

If the sand grain were to become stuck in the point bar at Great Bend, it most likely would not have the misfortune of being trapped for hundreds of millions of years like its ancient relatives. The chances are good that within a day or a few months, a storm would swell the river and flush it onward. The drainage area of the Susquehanna is so large that it frequently receives huge amounts of water from storms: it is one of the most flood-prone rivers in the country. In June 1972, Hurricane Agnes became a tropical storm and stalled for twenty-four hours over the region, pouring torrential rains into the drainage area. Over 500,000 sandbags were needed around Wilkes-Barre, Pennsylvania, where the water rose nearly 6 meters (19 ft) above flood level. The water flow in the Susquehanna tripled, and more than five million tons (200,000 dump truck loads) of suspended sediment surged past Harrisburg, Pennsylvania, in the space of a few days, several times the average for an entire year. No estimate has been made of the additional bed load hurtling along under this torrent. Stand beside a river in flood and you can *hear* the bed load in motion, the rumbling and clunking sounds of boulders tumbling along the riverbed. Major floods in 1996 were followed by record flows in early 1998, but a year later water flow was near the record low, less than half the normal rate. It is a dynamic river, the Susquehanna, and at any given time, our sand grain could be hurtling downstream or trapped in a sandbar, waiting for the next, inevitable flood to change the shape of the river.

RESPONSES TO CHANGE

A sand grain journeying down a river is a player in a great game of competing influences, as small as an eddy in an ever-changing channel and as vast as the rise and fall of sea level. The most dramatic shape-shifting that rivers display occurs along the bends. But why do rivers meander? Why not take the shortest, straightest path to the sea? Clearly, the shapes of resistant bedrocks across which the river must flow have a primary influence. If the softer rock formations, the ones that the river can most easily exploit, have a curved shape, then often the river will follow these shapes—but not always, as we shall see farther down the Susquehanna, where it appears to completely ignore the underlying geology. Where there is no bedrock influence, a river will meander regardless—and once formed, meanders are self-perpetuating, constantly exaggerating their shape.

But how do meanders form in the first place? The answer is that we don't know. We have some basic principles and experimental results, but the physics of fluid flow and sediment behavior is too complex. Even Einstein (presumably before advising his son to avoid the topic) turned his great mind to the formation of meanders, reporting a thought experiment on tea leaves collecting at the bottom of a cup whose contents had been stirred. The flow of water around the slightest bend sets up a complex of internal currents, each reacting with the roughness of the sediment on the base and the walls of the channel, modifying that sediment and, in turn, being modified itself. The smallest change will set off a whole series of essentially unpredictable further changes that feed back into the dynamics of the system. In experiments like those conducted in the Jurassic tank, a perfectly straight channel at the start will soon develop irregularities, which turn into bends, which develop into meanders. Perhaps the movement of a single sand grain is enough to trigger large-scale changes.

It almost seems as if rivers *like* to meander, to take the leisurely scenic route to the sea. Meanders represent a kind of self-regulating tool for a river. On the face of it, this seems a strange statement: surely a river is just a river, with no underlying game to play? But step back and look at the grand scale. Take a map of a river and run a piece of string along its winding course; then straighten out the string and draw a graph of the river's elevation versus the distance from its source—the profile of the river. For any river at any given time, there are two points that are externally fixed: sea level and the elevation of the source of the river in the mountains. Between these two points, the river is free to behave in any number of ways. For the river, sea level is the primary *base level,* its ultimate terminus, below which no part of the river can drop (to avoid the impossible feat of flowing uphill). Base level is a concept universally agreed on and at the same time hotly debated, but the principle is clear: change a river's base level by raising or dropping sea level, and the river must change its behavior. If base level drops, the river must erode, cutting downward to reflect this. If base level rises, the river must build up its profile by depositing sediment. A change in base level changes the gradient of the river's profile, requiring the river to make changes in order to smooth out that profile. Changing the geometry of meanders changes the length of the river and therefore its profile, and thus the river responds to changes in base level.

Changes in sea level, then, have far-reaching consequences; as it rises or falls, the effects are felt not just at the coast, but for a long way upriver. As we consider the consequences of the current rise in sea level, it's not just the shorelines we need to worry about.

In addition to changes in sea level, natural and human-induced base-level changes also take place locally along a river's course. Dams, levees, and artificial channels all create feedback into the larger architecture of a river. The river responds to even the slightest change, cutting down or building up, eroding or depositing sediment. Not only is our sand grain at the mercy of storms and floods, and the slow rise and fall of the Earth's oceans, but it also feels the effects of a world of minute local changes.

THROUGH THE MOUNTAINS

If the flow and capacity of the Susquehanna permit it, our sand grain tumbles past Great Bend, takes a sweeping trip back through New York State, past Binghamton, where recreational beaches used to line the river before the coming of industry and pollution. The sand grain reenters Pennsylvania, tumbling southeastward through meander after meander before turning right toward Wilkes-Barre. The flow of the river—and its sediment load—is ever increasing as it is joined by great tributaries, some of which begin only a short distance from Great Bend but take a more direct route to meet up with the Susquehanna here.

The right turn before Wilkes-Barre reflects a profound change in the continent beneath. Look on any map—satellite (Plate 6), topographical, or geological—and it appears as if a celestial artist has covered the land here in a sinuous, elongated calligraphy. These designs mark a distinctly different setting than that of earlier stages on our sand grain's journey. Geologists divide up the chapters of the Earth's past according to the sequence of rocks that tell the story, each one a *formation* that can be tracked and described and given a local name, and whose surface extent can be drawn on a map. Our sand grain's imprisoned colleagues back at Great Bend form part of the Catskill Formation, a gigantic pile of river and marine sediment, components of which can be easily examined in the sidewalks and curbstones of New York. (We shall see the Catskill Formation again in chapter 7.)

The mountains that provided the source of the floods of sediment of the Catskill Formation were heaved up in just one episode of the slow movement of continents and digestion of oceans that culminated, nearly 300 million years ago, in the welding of the continents into a single supercontinental entity, Pangea ("all Earth," in ancient Greek). The titanic collisions that occurred treated the edge of the North American continent like putty, slowly but relentlessly bulldozing the rocks into the great wrinkles and folds (now subdued by erosion) that create the sinuous patterns of our celestial calligrapher's work. This is the Valley and Ridge Province of the Appalachians, and it is engaged in an ancient battle with the Susquehanna. The collision disinterred almost every kind of rock from almost every period of the continent's past, resistant rocks forming the ridges, softer and more easily erodable ones the valleys. It is when the Susquehanna encounters the first of these northeast-trending structures that it turns sharply right and enters the Wyoming Valley, treasured by its Iroquois inhabitants and seen by its earliest Western visitors as a quintessentially American paradise. The valley is a geological as well as a topographical depression and is filled with rocks of the Llewellyn Formation, sands, silts, and, sandwiched among them, seams of coal, originally the economic foundation of the region. The Appalachian coals are high-grade anthracite, first used in a blacksmith's in Wilkes-Barre in 1769. The Susquehanna has provided a highway for the coal industry, allowing boat transport and carving a route for the railway. And it has caused tragedy. The most accessible coal seams are close to the surface, and in January 1959 miners in the Knox Mine, just upstream from Wilkes-Barre, tunneled far beyond the limits of safety, until only a small thickness of fragile rock separated them from the bed of the flooding and ice-filled Susquehanna. With a sound like thunder, twelve billion gallons of water and sediment poured into the mine and twelve miners lost their lives. Mines were closed and the tragedy led to the end of mining in the Wyoming Valley.

CHANNEL BARS

The Knox Mine disaster took place next to Wintermoot Island, a long bank of sand and gravel in the middle of the river. Downstream, other islands break up the main channel, one of them helpfully facilitating a bridge crossing. But what are these islands doing in the middle of the river, where the current should be fastest, sweeping the sediment along? These are *channel bars* (Figure 22), and their origins are no better understood than meanders. Something in the shape of the channel bed, some small changes in roughness, in frictional forces, in turbulence, splits the current and slows it down, causing the deposition of sand; the process continues and the island grows, emerging above normal water levels and building up downstream. Does the channel bar grow from the ripples and wave forms sculpted into the sandy bed of the river by the current? Or does a split in the water flow result in two coalescing point bars? The internal structure of channel bars, like that of point bars, is similarly characterized by cross-bedding. They can segment the river into so many different channels that its pattern is described as *braided*.

Channel bars can be large and take up much of the channel, hindering navigation but aiding river crossing and providing habitable space. They are fragile and unstable, however, constantly shifting size and position, steadily marching down the river, disappearing and reappearing during floods and under the attack of ice. Tropical storm Agnes, for example, completely changed many of the Susquehanna's islands. Historic maps and paintings of the river, compared to its appearance today, graphically illustrate these changes.

One dramatic result of the 1811 New Madrid earthquakes was the appearance and disappearance of channel bars along the Mississippi. But regardless of seismic events, the naturally peripatetic nature of Mississippi sandbars can cause problems—even legal ones. A channel bar was the subject of a long-running boundary dispute between the states of Mississippi and Louisiana. Stack Island is a large, uninhabited channel bar, sometimes above water and at other times not. In the early nineteenth

FIGURE 22. A river, braided with channel bars. (Photo by author)

century, it was close to the Mississippi side of the river and the state line ran down
the main channel to the west of the island. Time and the mighty Mississippi moved
the island across and down the river until the navigable channel lay on its eastern
side. Since the law states that, by the "rule of thalweg" (*thalweg* being another word
for a main navigable channel), a state boundary must run along that channel, the
State of Mississippi felt that the State of Louisiana had performed a land grab and
sought legal retribution. After many years, the case went all the way to the Supreme
Court, which came to a final decision in 1995. Louisiana had conceded that in 1881
there had indeed been a Stack Island and that it had belonged to Mississippi, but
insisted that "two years later, in 1883, Stack Island washed away and was replaced by
mere alluvial deposits, which at various times over the last 100 years were not sufficient
in size or stability to be deemed an island." However, after lengthy geomorpholog-
ical rumination, the Supreme Court could find no documented or sedimentological evidence for the claimed disappearance, invoked the "island exception to the rule

of thalweg," and allocated Stack Island back to the State of Mississippi (*Louisiana v. Mississippi* [121ORIG], 516 U.S. 22 [1995]).

Out of the Wyoming Valley, the river (and, with any luck, our sand grain) makes a sharp left turn and then a right, across into the next valley. Along its way, it dissects an exotic geological lexicon—the Onandaga Formation, the Mauch Chunk, the Wills Creek, and the Tuscarora—traveling past the towns of Shickshinny, Wapwallopen, and Catawissa. At Sunbury, the river joins forces with a great tributary, the West Branch. The flow increases, the channel widens, and the valley broadens. In the nineteenth century, there was a dam here to permit ferry crossings, but it was destroyed by ice in 1904. However, if our sand grain attempts the passage during the summer today, it faces another impediment—the world's longest inflatable dam, used to create an area of water for recreational purposes (and which interferes with the local base level). If our sand grain doesn't make it while the dam is inflated, it will only be a short wait until the air is let out and the winter floods arrive.

AN OLD RIVER

A little farther down, the river faces further challenges from the structure of the Valley and Ridge Province—and overcomes them by doing something very unexpected. Shortly after the confluence with its northwestern tributary, it encounters Mahantango Mountain, one of the great ridges formed from the resistant sandstones of the Mauch Chunk Formation. We might expect the river to be forced to find a way to wind around the end of the ridge to the west, but it doesn't: it simply carves its way straight through the mountain to form one of the great "water gaps" of the Appalachians. A little farther south and it does the same thing at Peters Mountain—and again at Blue Mountain, just north of Harrisburg. How it has come to do this is, again, the subject of some conjecture and dispute. The mountainous topography that we see today had, for a long period of geologic time, been covered in layers and layers of younger rocks, across which the old Susquehanna had flowed. The river stripped away these rocks and cut down through them, eventually excavating and revealing the old structures of the Appalachians as the topography evolved. Being already established in its course, the river simply kept cutting down, like a knife beginning to slice the filling after penetrating the top crust of a pie. While we have yet to pin down exactly how the river accomplished this remarkable feat, it is clear that the Susquehanna is a very old river. But how old? We do know that, in its games with the comings and goings of ice sheets and changes in sea level, the river eroded far deeper valleys, now filled with sediment, than those we see today; the Susquehanna's ancestors were certainly carving their valleys and transporting our sand grain's predecessors a couple of million years ago, or even further in the past.

As it flows past Harrisburg, the river takes on a different appearance. It curves gently, but without meandering, and it is festooned with hundreds of channel bars; point bars develop at some of the gentle curves, particularly where tributaries join. The river is split into so many different channels that it is braided, a sign of large amounts of sediment being transported (Figure 22). The overall channel is very wide, nearly two kilometers (over a mile) at Harrisburg, and the valley floor, while still constrained by the mountains, is in places broad and flat—a floodplain. This is, of course, a very modest floodplain—that of the Mississippi is 200 kilometers (125 mi) wide in places—but the processes are the same. As meanders migrate sideways, carrying successive point bars with them, as channel bars follow that migration, and as the channel floods, pouring out sediment onto the flat land around, so the floodplain is formed (Figure 20). As time passes, the channel migrates back and forth across its floodplain, building it up layer by layer. The floodplain is, in the first instance, ideal for human settlement and transportation—Harrisburg, for example, is built on a floodplain—but it is, by definition, vulnerable to floods. If base level drops, however, then the river cuts down into its floodplain and leaves broad benches, or terraces, of the old floodplain at higher levels, where they are more secure above the new flood level. Alluvial terraces, built of sand and other sediments, have for millennia provided the platforms for human development around the world.

Below Harrisburg, the river passes through another of the great anthracite mining districts of Pennsylvania. A steel mill was built in Harrisburg in 1866 and supplied by local coal. For decades, the coal was broken up and cleaned in the local streams, which, together with debris from mine tips, created a vast input of coal fragments and dust into the river. For decades, this created an industry downstream where Susquehanna sediment was dredged and screened, the sand separated out and the coal taken to Harrisburg for sale. Around 250,000 tons of coal a year were recovered this way, and the channel bars in the river today are full of it—commercial coal dredging has continued into recent times far downstream. Mines were—and are—also the origin of acid waters draining into the river. While little physical change has happened to this part of the river as a result of mining, agriculture, and other human sources, its chemistry can be distinctly unpleasant. Major efforts are underway to change this—the Wyoming Valley watershed and the Susquehanna down to Sunbury have been designated as an American Heritage River.

As our sand grain approaches the Maryland state line, it has more obstacles to overcome in addition to sandbar traps: three dams. But these dams were built in the first half of the last century; two of the reservoirs are now filled with sediment, and one is nearly so. Soon after a dam is built, the deep, empty waters behind it effectively entomb large amounts of sediment. But the accumulation of sand and mud eventually results in shallow water, bringing the bed within the reach of storms and floods. Therefore, if our sand grain does get stuck, it is likely to be rescued by the periodic great floods of the Susquehanna. During the 1996 floods, 15 million tons (600,000 dump truck loads) of sediment were flushed through, and scoured out of, the reservoirs—and so the sand grain has a good chance of making it.

Not far below the dams, the Susquehanna empties abruptly into the Chesapeake Bay. The river provides around 50 percent of all the freshwater entering the bay and, *on average,* around 1.8 million tons of suspended sediment each year, perhaps 10 percent of which is sand. In addition, the river's coarser bed load is flushed out into the bay. Once the reservoirs have been filled with sediment, the river's load will more than double, with consequences that will require some thought and planning. Even today, deforestation, agriculture, mining, and other human disturbances have dramatically increased the rate of sediment transport: when Captain John Smith first explored the lower reaches of the Susquehanna, it was probably transporting half the sediment it does today.

AN ARM EXTENDED IN WELCOME

In *The Pioneers,* James Fenimore Cooper described the Susquehanna as "a river to which the Atlantic herself has extended an arm in welcome." That arm, the Chesapeake Bay, has been extending for some time and continues to reach up the river. Eighteen thousand years ago, the Susquehanna would have had much farther to run beyond Havre de Grace, Maryland, its present entry point into the bay. The low sea level during the last of the great glacial advances meant that the shoreline was far to the east, and the channel cut by the ancient river as it worked toward this lowered base level is detectable below the sediments of the bay floor. Sea level—base level—rose with the melting of the ice, and the ocean invaded the bay. The river backed up, depositing its load in response to the dramatic changes. Sea level continues to rise today at the rate of around 4 millimeters (⅙ in) every year, and the river continues to respond.

The Chesapeake Bay is an estuary, a river valley drowned by the sea, and it is the largest in the United States. Because of the effects of glaciation and the resulting changes in the level of the land and the sea, all the rivers of the northeast coast find themselves emptying into estuaries—there is no opportunity to build out into the sea in the way that the great deltas of the Mississippi and other rivers do. Estuaries are the stage for a constant interaction between river and sea, between freshwater and salt water. Twice a day, the tide floods up the bay and ebbs out again, acting like a gigantic pump for water and sediment. Tidal currents set up in the bay are strong, but their effects vary: in some estuaries and at some times of year, the flood tide dominates, at others the ebb, and circulation is complex. Estuaries are dynamic places, but they are also of historic commercial importance the world over. They are, after all, sheltered harbors, which makes them hospitable entrances to the continents and their great rivers. Why did the first exploration of the interior and much of the first settlement of what would become the United States spring out of the Chesapeake Bay?

As Joseph Conrad wrote in *The Mirror of the Sea:*

> The estuaries of rivers appeal strongly to an adventurous imagination. This appeal is not always a charm, for there are estuaries of a particularly dispiriting ugliness: lowlands, mud-flats, or perhaps barren sandhills without beauty of form or amenity of aspect, covered with a shabby and scanty vegetation conveying the impression of poverty and uselessness. Sometimes such an ugliness is merely a repulsive mask. A river whose estuary resembles a breach in a sand rampart may flow through a most fertile country. But all the estuaries of great rivers have their fascination, the attrac-

tiveness of an open portal. . . . That road open to enterprise and courage invites the explorer of coasts to new efforts towards the fulfilment of great expectations.

Conrad went on to describe the precautions needed around the sandbanks of the Thames Estuary. Sand grains are the ground troops in the daily battle between river and tide, and they make continuous assaults and retreats, massing again in different places for the next phase of the campaign. The Thames Estuary is strewn with shifting sandbanks whose movements are constantly monitored. One is particularly closely watched. In August 1944, the USS *Richard Montgomery* arrived in the estuary. She was a Liberty ship, laden with a cargo of bombs for the campaign in Europe, but she never got to unload her cargo as planned: misdirected to her berth, she ran aground on a sandbank, stuck fast, and began to break up. Some of the cargo was salvaged, but the efforts had to be abandoned. The *Richard Montgomery* remains there today, a literal time bomb. On board still (or buried in the sand around the wreck) are 3,000 tons of munitions, equivalent to 1,400 tons of TNT. If they exploded, it would be one of the biggest non-nuclear blasts in history—and close to population centers, docks, refineries, and gas terminals. The wreck lies in the center of the shipping lanes in the estuary, marked by a buoy and governed by strict rules of avoidance.

The Chesapeake Bay contains no such drama, but it has seen many changes over historical times. About halfway down the bay and close to its center is the Sharp's Island Light. The structure, which looks like the Leaning Tower of Pisa, is now in 3 meters (10 ft) of water, and there is no Sharp's Island. In the seventeenth century, there was an island there, originally mapped by John Smith; at 3.5 square kilometers (900 acres) in size, it came to support a thriving farming community. The current lighthouse, built in 1882, is the last of a series designed to warn of the shoals and sandbanks in the area; all succumbed to shifting sands and the destructive power of winter ice floes. The island itself still appeared on charts at the beginning of the last century.

In the Chesapeake Bay, like most estuaries, it is the sea that is winning the sand battle with the rivers—most of the sand in the bay is swept in by wave and tide from the Atlantic and strewn around the estuary. As soon as our sand grain hits the bay, it meets the zone where freshwater and salt water, pumped in by the tides, mix—the *EMT,* or estuarine maximum turbidity zone. Unfortunately for our sand grain, the EMT is a very effective sediment trap, and much of the Susquehanna's load is dumped soon after entering the bay. Very little river sand makes it very far into the estuary, but some does. Perhaps during a major flood or storm, a particularly high tide or ferocious currents, our sand grain will be picked up and flushed out into the ocean, to join the trillions of its colleagues already on their coastal journey south toward the Outer Banks. It might even enjoy, like Paddle-to-the-Sea, some human help and make it by boat or in some fishing tackle, or in the sole of someone's boot. Let's assume it does.

5

Moving On

Waves, Tides, and Storms

The same regions do not remain always sea or always land,
but all change their condition in the course of time.

Aristotle, "Meteorologica"

I with my hammer pounding evermore
The rocky coast, smite Andes into dust,
Strewing my bed, and, in another age,
Rebuild a continent of better men.

Ralph Waldo Emerson, "Seashore"

A TALE OF TWO LIGHTHOUSES

If you had happened to be enjoying a fine day in late June 1999 by taking a walk
along the beaches of Cape Hatteras, on the Outer Banks of North Carolina, you
would have thought that your mind was playing tricks. You might have dismissed
it as an illusion of some sort and sat on the beach, keeping an eye on your dog and
sifting sand grains through your fingers—any one of which might have been our
sand grain resting on its journey southward after its escape from the Chesapeake.
But the sensation of uneasiness would have remained, and you would have inevitably
turned back to look again at the huge, candy-striped lighthouse—and yes, it had
moved.

This would not have been a trick or an illusion, a sign of your deteriorating fac-
ulties. The Cape Hatteras lighthouse, the highest in the country at over 60 meters
(208 ft) tall and a historic icon, was on a slow, 900-meter (2,900 ft) journey from
where it had stood for 129 years to a new, safer location. When it was built in 1870,
it stood 500 meters (1,500 ft) from the shore. But the same treachery of the ocean

that required the lighthouse to warn ships of offshore shoals had threatened the structure itself. By 1935, the waves were crashing only 30 meters (100 ft) from its base. The beach had gone, carried away not just by the regular storms of the Atlantic, but also by daily wear and tear, washed away by the same processes that move the sand between your toes. The cumulative effect of those processes is that Hatteras Island in its entirety is continuously migrating westward.

To move a lighthouse is no trivial task, and many different alternative engineering solutions had been tried over the decades. Since the 1930s, huge lengths of massive steel sheets had been driven into the shoreline to form groins, walls against the sea. The groins were extended and repaired, but time after time storms found their way around them to slam into the defenseless sands and move them on again. In 1966, fourteen thousand dump truck loads of sand were pumped from neighboring Pamlico onto the beach in front of the lighthouse, but the sand was fine, puny stuff, easily dismissed by the waves, and dismiss it they did. In 1973, in a further attempt at what, in the trade, is referred to as *beach nourishment,* another fifty-nine thousand loads of sand were moved from the cape to the lighthouse defenses, only to be swept away within a few years. By 1980, the sea was once again threatening the lighthouse, and sandbags, rubble, and rubber mats were dropped offshore. None had any lasting effect in the face of the stubborn aggression of the ocean. In 1988, the decision was made to retreat, to move the lighthouse, lock, stock, and lantern, inland. The structure, weighing close to 4,000 tons, was jacked up off its foundation in June 1999 and moved on tracks at around 2 meters (6 ft) per hour over a period of twenty-three days. The lighthouse now stands roughly the same distance from the sea as it did when it was first built. The waves continue their action— as they have, of course, been doing for millennia.

But the forces of nature can be constructive as well as destructive, giving some lighthouses a different problem.

Thirteenth-century France was not what it is today; it was fragmented and torn by conflicts between counts and kings, Spain, Toulouse, and the Holy Roman emperor. Having extended their domains southward through crusades against the peaceable residents of Languedoc, the kings of France needed a Mediterranean port for commerce and further military ambitions. Around 1240, Louis IX decided that an area of lagoons and shifting sandbanks in the Camargue, where the Rhône empties into the Mediterranean, would be ideal. The construction of what would become the extraordinary walled town of Aigues-Mortes began. In medieval times, a tower was generally the way to start a town, and the Grosse Tour du Roi of Aigues-Mortes was completed in 1248.

Although it later earned infamy as an unspeakably cruel prison for Huguenot women, at the time of its construction this tower served a different purpose. In addition to being the anchor of the city's fortifications, the tower functioned as a lighthouse for the developing port. The port, in turn, was linked inland via canals and rivers to Arles and Montpellier and flourished, rather like France today, through a combination of enterprise and regulation. It also served as the starting point for Louis IX's ill-fated foreign crusading initiatives.

The king's luck at selecting the location of a port was no better than that of his military escapades. The Rhône delta is a constantly shifting environment: vast volumes of sand and silt are carried down by the river to the sea, which then takes over the action by shifting everything around. The river moves, the sandbanks move, and the lagoons move, measurably, every year. The port slowly but inexorably filled up with sand, and no amount of medieval dredging could provide a sustainable solution. From the early fourteenth century, the requirement to use this increasingly shallow port was the cause of escalating complaints by shipping merchants. Politics conspired, and when Marseille was acquired by France in 1481, the dominance of Aigues-Mortes was over. The lighthouse that had guided the trade of spices from the East, wool from England, gourmet produce from Champagne, and salt from the local marshes had no further role to play.

Two different lighthouses, two different problems: one, too little sand, the other, too much. The coastal rules of supply and demand of sand operate on a grand scale—and are immensely complicated.

PLAYGROUNDS

Rivers are arteries, gathering and transporting sand to the sea; the coasts and the oceans are sporting arenas for some spectacular events. The players are many and diverse: waves, tides, currents, winds, storms, gravity, topography, geology, history—and humans. For a sand grain entering the game (such as ours from the Susquehanna), the players make the rules: where the grain may go and where it may not, how it will get there, how long it will take, how long it must remain in any one place, and in what sort of company it will travel.

Like a river, an ocean is an engine, conspiring with the atmosphere to do work, often on a grand and awe-inspiring scale. This engine covers around 71 percent of the planet's surface, and it does much of its work in its game with sand along the coasts. The game is a contest between construction and destruction, deposition and erosion. But the contest is uneven—the waves and currents exert tens of thousands of times more energy than it would take to keep all the sediment along the world's coasts constantly on the move. The result is that, while the ocean may allow occasional pauses during which new beaches and sandbars are formed (and occasionally preserved for posterity), the waves and currents almost always win in the end.

In order to be able to follow the game, to track, map, and—occasionally successfully—predict the movement of sand around our coasts, we need to understand the arena. Shorelines are transient things, their position depending on sea level at any given time. And sea level itself depends on several different factors. First of all, *absolute* sea level is a global measure and simply a result of how much water there is in the world's oceans, that volume depending on how much of the planet's water is locked up in continental ice sheets and the atmosphere, and on the temperature of the water (the warmer it is, the more volume it occupies). Absolute sea level fluctuates constantly, and in the past, as we have seen during the advances of the ice sheets, it has been dramatically lower than it is today. *Relative* sea level records the position of the sea at any particular place and reflects not just absolute sea level, but the nature of the land and its intersection with the sea at that point. The land may be rising or subsiding as a result of continental-scale or local movements; sediments may be building up or erosion taking place. Changes in relative sea level affect parts of a particular coast; changes in absolute sea level have worldwide effects. It's like being in a bathtub—even if the absolute water level stays the same, local water level might be at your chin or your chest, depending on your position.

Globally, absolute sea level is rising slowly. The rate may seem imperceptible, but it has been far from an incidental factor in human history. Our British ancestors made a good life for themselves on the floor of what is now the North Sea; the ice may have made for a cold climate, but it resulted in large areas of habitable land.

As the ice melted, that land was so flat that it was inundated rapidly—the North Sea dwellers had to retreat from a shoreline encroaching by meters every year. Today, the Chesapeake Bay continues to invade the Susquehanna Valley, but during the glacial periods, disproportionately large areas of land were exposed, and the ancient Susquehanna had a much longer way to go to reach the sea. But the shorelines as we see them today only reflect the current stage of the contest between land and sea, and bear no relation to the true boundary between continents and oceans.

LANDSCAPES BENEATH THE SEAS

Plate 7 shows the topography of much of the Americas and the western Atlantic with the plug pulled and the ocean emptied. Shorelines as we know them are obvious—but they are not the edges of the continents. The true structural edges, beyond which there is a drop to the deep ocean floor, are often a long way beyond the shore—up to 1,500 kilometers (900 mi) distant. The flooded margins of the continents, covered by relatively shallow seas, are the continental shelves (appearing in light blue), and it is in these arenas that much of the action takes place, where many of the hundreds of millions of dump truck loads of sediment brought down to the coast each year end up. Some of this sediment, as we shall see, escapes to the deep oceans, but, relatively speaking, not much.

During the advances of the ice, absolute sea level was close to the edge of the shelves; the extended journey that the old Susquehanna had to make can be seen from the extent of the continental shelf—above sea level in glacial times—beyond the mouth of the Chesapeake. The apparently precipitous drop to the deep ocean takes place across the continental *slopes*, in reality not cliffs for most of their extent, but they do represent in total area the steepest topographical zones on the planet. The topography of the oceans is as diverse as, and on a larger scale than, that of the continents, and it is a realm about which daily scientific revelations continue to be made.

It is only relatively recently that we have even begun to understand the complexity of the ocean. On December 21, 1872, fourteen years after Charles Darwin had published his revolutionary ideas, a small converted British naval vessel slipped out of Portsmouth, beginning a voyage that would last three and a half years and revolutionize our understanding of the oceans. HMS *Challenger* (after which the space shuttle was named) had been converted into a research vessel. Her guns had been replaced by laboratories and her munitions stores by over 400 kilometers (250 mi) of rope and wire for surveying and sampling the ocean floor. The voyage covered 130,000 kilometers (80,000 mi) and cost more than $20 million in today's money. The incentives were many, but two stood out: the needs to plan for global communication via undersea cables and to test the long-standing belief that the deep oceans were lifeless wastes. The latter was dramatically proved wrong, and the former accomplished by large numbers of soundings that first demonstrated the topographical grandeur of the ocean floors. In waters off the Pacific Mariana Islands, a depth was recorded of 8,200 meters (27,000 ft). What became known as the "Challenger Deep" remains the world-record holder today, its greatest known depth later measured at close to 11,000 meters (36,000 ft). Ocean chemistry, temperatures, and huge amounts of other data were collected and, with the publication of the fifty-volume report, oceanography was born. However, it was not until 1957 that the first comprehensive map of the Atlantic Ocean floor, a forerunner of the one shown in Plate 7, became available, providing the underpinning for what would become the ideas of moving continents and plate tectonics.

Today, ever more sophisticated technology allows us to do increasingly detailed imaging of the ocean arena and the activities within it. And while, in many ways, there remain more questions than answers, our understanding of the oceans tells us a great deal about what is happening on the edges of the continents. The Earth's crust beneath the oceans is geologically young, thin, dense, and, except for a veneer of sediment, composed of igneous rock. The crust beneath the continents is ancient and thick, made up of a huge variety of generally low-density materials. The submarine continental slopes mark the weld between these two very different crustal types; if the Atlantic were removed, the shapes of the Americas and Europe and Africa *along the outer edges of the continental shelves,* not along their shorelines, would fit together like the pieces of a rough jigsaw puzzle. The continents have been, and are being, moved apart by the daily creation of new oceanic crust, produced by the great ridge of volcanic mountains curving down the center of the Atlantic (the Mid-Atlantic Ridge, identified in our map by its paler blue coloring and rugged topography). The continents are moving apart with little disturbance, except for slow subsidence along their edges as they move away from the deep-seated molten upwelling at the mid-ocean ridge. This subsidence allows the accumulation of thick blankets of sediment on the continental shelves.

The shelves and coasts on either side of the Atlantic are distinct from those where the Earth is less quiet. On the west side of South America, the shelf is narrow and bordered by a deep trench, similar to the one the *Challenger* found. Here, to accommodate the westward movement of the continent, the oceanic crust and its substrate are foundering into the Earth's molten interior: as the Atlantic grows, the Pacific shrinks. That process, marked by trenches in the ocean floor, is *subduction.* The result, demonstrated by the "ring of fire" that encircles the Pacific Ocean, is crustal turbulence on a grand scale: earthquakes and volcanoes, rapid uplift and subsidence. Coasts and shelves along the edges of continents where subduction is taking place differ greatly from their passive equivalents along the Atlantic. Most coastlines in the world fall into one of these two categories, but some, such as California's, are developed where plates are lurching sideways, past each other, creating the fabled diversity of California's shores.

Such are the global forces that influence the rules of the game for the sand on the edges of continents. Other forces, as we will see, are more local. The rule book is large. As every grain of sand is unique, so is every segment of coast.

WELL-STOCKED SHELVES: SUPPLY AND DEMAND

Sand is being supplied continuously to the continental shelf, and waves, tides, and currents are constantly demanding more recruits to move around. Our sand grain finds itself in an army of its companions that is constantly being mobilized southward along the Atlantic coast. The next battlefield is the long chain of islands—barrier islands—that form the Outer Banks of North Carolina. Much of the Atlantic coast of the United States is fringed by barrier islands, but this stretch is nearly continuous—and very famous. Called *barrier islands* for obvious reasons, they shelter the coast behind them, protecting estuaries and lagoons, breaking the power of the ocean storms. But in doing so, they are at the mercy of those forces themselves, and they are fragile; as a result, they are constantly, sometimes catastrophically, on the move.

As any sailor knows, making charts of a coast is a necessary but ultimately thankless task, one that has to be repeated almost as soon as it's completed. Nature abhors being tied down and demonstrates this nowhere as emphatically as along a coast, and a barrier island coast in particular. We have more than sufficient history to see changes in the charts, as was the case with Sharp's Island in the Chesapeake Bay. Providing shelter and access to the interior, the Outer Banks played a role in early settlement and exploration, but one of the reasons why the first English settlement, the "lost colony" of Roanoke Island, remains lost today is that Roanoke Island has moved.

Close to Roanoke Island, the Oregon Inlet provides a key break across the Outer Banks for navigation, but the inlet, and the islands on either side, refuse to stay put (see Plate 6). The U.S. Army Corps of Engineers has been examining for years what strategies are possible—and within its control—for keeping it open, waging a long-standing campaign here and elsewhere in the Outer Banks. In 1828, the Corps was commissioned by Congress to acquire a dredging machine and open up Ocracoke Inlet, farther south from Oregon Inlet and equally fickle. After wearing out several dredges, building various jetties that would be destroyed by storms, and spending $130,000, the Corps gave up in 1837—the sand had won. Our capacity for moving sand and our understanding of how barrier islands work may have improved enormously since then, but long-term solutions are no easier to manage. The underlying problem, which we shall return to, is that we have developed an unnatural desire to live in places that we shouldn't.

Defeating the Army Corps of Engineers wasn't the first victory for the sands of Ocracoke. In June 1718, Edward Teach, otherwise known as Blackbeard, pirate and tyrant of the eastern seaboard, had almost outmaneuvered Lieutenant Robert Maynard of the Royal Navy among the shifting shoals off Blackbeard's sheltered camp on Ocracoke Island. But a sandbar outmaneuvered him, and his ship, the *Adventure,* stuck fast. A gruesome battle ensued, Blackbeard was decapitated, and legend has it that his headless body swam several times around the ship before disappearing. Earlier that year, Blackbeard had, some say intentionally, run his flagship, the *Queen Anne's Revenge,* aground a little farther south, in Beaufort inlet. The storms of 1996 partially scoured a wreck from the sand, claimed to be the pirate's flagship.

Barrier islands require a substantial and ongoing supply of sand. For the Outer Banks, this is no problem, for the same glaciers that brought our sand grain to the headwaters of the Susquehanna abandoned, in their retreat, huge quantities of their cargo of sediment on the continental shelf—at that time above sea level. As the sea rose and the shelf was inundated, the encroaching ocean had piles of sand and gravel, clay and boulders, to play with. Rivers bring down further sediment each day, but it is still the glacial refuse that provides most of the grist to the mill of the Atlantic coast today. But why was the sand sculpted into this extraordinary string of islands, with the distinctive "elbows" at Cape Hatteras and Cape Lookout? Yet again, we don't have the complete answer—the islands move at a faster rate than our understanding of them, and their origin has been the subject of intense debate and different schools of thought for more than a hundred years. Some may have formed through the growth of offshore sandbars, the bars interfering with wave and current movements and thus causing more sand to be deposited until the bars rose above water level. But what specified the initial sandbar location? Others may have formed by the elongation of spits of sand streaming downcurrent from headlands. It seems quite likely that the Outer Banks resulted from the westward-encroaching ocean reworking old beaches and sand dune systems from the ice ages—the islands are certainly still migrating westward.

The east coast of the United States and the Gulf of Mexico contain the longest barrier island systems in the world. Less than 15 percent of the coasts of the world sport barrier islands, and the Atlantic coast has 23 percent of them. They behave in very distinctive ways, responding dramatically to waves, tides, and storms. Every wave that hits a beach is shifting tons of sand. If the waves strike the beach straight on, the sand is shifted up—and down—the beach, but if the waves strike at an angle to the shore, then sand is transported along the beach with every wave. This "longshore drift" is characteristic of most beaches, including those of the Outer Banks, where sand is marched southward down the shore. Where there is a break in the island chain, such as at Oregon Inlet, the net effect is that the shoreline to the north of the inlet grows southward and that on the southern side retreats through erosion. This effect is complicated by the tides: flood tides transport sediment through the inlet from the sea, and ebb tides have the reverse effect, building out deltas with each flush of sediment. The net result is the highly complex, large-scale, and seasonally changing movement of sand around the inlet—and a challenge for the Army Corps of Engineers.

Natural breaks in barrier islands not only move, but also appear and disappear almost instantly. The islands respond to storms with dramatic and rapid change. In September 2004, Hurricane Ivan generated the most extreme storm waves ever measured, more than 27 meters (90 ft) in height, devastating the Caribbean. The hurricane then slammed into the U.S. Gulf Coast, causing widespread death and destruction. Figure 23 shows "before" and "after" photographs of the state park at Gulf Shores, Alabama. Roads and parking lots have completely disappeared under blankets of sand, the fishing pier has been fragmented, dunes have been destroyed, and a new break in the barrier island has formed.

Barrier islands vary in form depending on the local balance of waves, tides, and sediment supply, but they typically share some basic features. The beach—the reason these islands are so popular for high-risk residence—faces the ocean, generally backed by a system of coastal sand dunes, some stabilized by vegetation, some actively on the move. The beaches and dunes are where most of the work of the ocean is expended, and behind the islands the natural energy is low (except during hurricanes). Marshes, tidal flats, and lagoons are home to an important variety of flora and fauna sheltered from the ocean. During major storms, however, this tranquil environment is threatened. Sand is torn from the beach and poured inland, covering the marshes and tidal flats and being dumped into the lagoon; the unrelenting force of water and sand carries destruction in its path.

The residents of islands in the Gulf of Mexico know this only too well. During Hurricane Ivan, much of the beach along Dauphin Island, off the coast of Alabama, was transported across the island, reducing its size and changing its shape. After Hurricane Katrina, in 2005, the beach was stripped and spread completely over the island as sheets of sand, known, unsurprisingly, as *washover fans.* Houses and entire streets disappeared.

Processes such as these result in barrier islands migrating wholesale toward the mainland—unless they are inhibited by our attempts at preservation, in which case they completely change their natural behavior. That behavior responds to the laws of supply and demand and operates on a *budget.* As with any budget, there

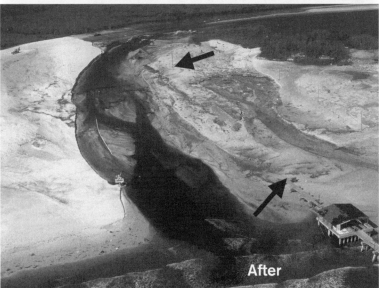

FIGURE 23. Aerial photographs of a section of the Alabama coast before and after Hurricane Ivan in 2004 (the ocean is in the foreground). The arrows mark the same points; note the buried parking lot. The barrier island has been breached and a new channel has broken through to connect the ocean with the lagoon (not visible in these photos). (Photos courtesy of U.S. Geological Survey)

are inputs and outlays: sand is added and removed. If there is a surplus, sand is deposited; if there is a deficit, erosion and land loss prevail. Understanding the sediment budget for a given location on the coast is key to defining both natural processes and the consequences of human interference, the budget measuring input of sediment versus removal, directions of transport, and volumes and variations with time of year. Sediment budgets for the world's beaches are particularly important, given their exposure and the concentration of human activity along them. Certainly, since the Romans, the beach holiday has been an institution—Cicero, Pliny, Tiberius, Pompey, and countless other intellectuals and emperors repaired to the beach for relaxation.

In September 1900, two brothers, bicycle mechanics from Ohio, arrived at a beach on the Outer Banks just up the coast from Oregon Inlet. They had come not for a holiday, but because the beach offered constant strong winds and long reaches of flat sand. They set to work not far from a small, desolate fishing village, an outpost named Kitty Hawk. Sand was both the key and the challenge, as Orville Wright remarked in a letter to his sister: "But the sand! The sand is the greatest thing in Kitty Hawk, and soon will be the only thing. . . . The sea has washed and the wind blown millions and millions of tons of sand up in heaps along the coast, completely covering houses and forest." The Wright brothers suffered setbacks from moving sand, but the wind blew, the gliders took off from the dunes of Kill Devil Hills, and on December 17, 1903, the world changed with the advent of manned, powered flight, takeoff and landing assisted by sand. The memorial to the Wright brothers on a Kill Devil dune has to be firmly shored up to remain in place.

AN ENDLESS DANCE

In their cultural history *The Beach: The History of Paradise on Earth,* Lena Lenček and Gideon Bosker provide a lyrical but geologically accurate description of their subject:

> The beach is not so much a distinct place as it is a set of relations among four elements: earth, water, wind, and sun. Partnered in an endless dance, these elements produce a staggering range of beaches, each subject to constant change, sometimes

rhythmical and cyclical, sometimes linear and catastrophic. If there is a single invariant played out on the boundary of land and sea, it is contained in the paradox of ceaseless metamorphoses, in the idea of immutable mutability. Minute by minute, hour by hour, each of the four constituents submits to the action of the others, and each, in turn, bends the others to its influence.

Where earth, water, and wind perform together, when the shape of the coast, the slope of the seabed, the supply of sand, the direction of the waves, and the strength of the tides work in just the right way, a beach is formed. Although the details of this process can cause any one of a wide range of beach types to form, and although there are details that we have yet to understand, there are some basic rules.

First, waves. Generated by friction between moving wind and the surface of the water, waves out in the ocean can vary significantly in size, both in wavelength (the distance between crests) and height. A wave is simply a moving form on the surface of the ocean—the water beneath the crest is rotating in the direction of wave movement, the water in the trough in the opposite direction. There is no net movement of water, only of the form of the surface. At increasing depth below each wave, the amount of water circulation decreases, until, at the depth of the *wave base,* there is no effect of the wave passing. Wave base is at a depth of half the wavelength, and so large waves have a much deeper influence. In the open ocean, there are waves with wavelengths of up to 600 meters (2,000 ft), and as these monsters roll in across the continental shelf, whose depth is typically less than 200 meters (650 ft), the entire seabed is affected. But smaller wavelengths are more common, and it is only when one of these waves reaches shallow water that it begins its dance with the sea floor. As the wave enters shallow water, there is less and less room for the movement of water beneath, and so the mass of water builds up vertically, forcing wholesale displacement of the water in the direction of the wave's travel; the wave shovels up sand from the bottom, begins to break, and crashes onto the beach (Figure 24). Thousands of tons of sand may be carried in by the wave, and thousands more moved as it breaks. Add in the effects of the tides, the direction of the breaking waves, the shape of the coastline, and the time of year, and sand is moved up the beach, down the beach, or along the beach. In such a dynamic setting, at any given time there can be net accumulation or erosion.

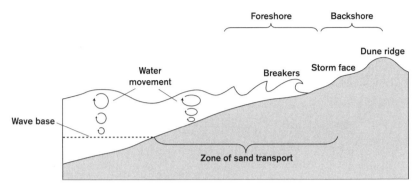

FIGURE 24. Breaking waves and a typical beach profile.

The slope and topography of the sea floor beneath the incoming waves have a strong effect on beach activity. Where the slope is steep, the wave heaves up, but the water at the base outruns the crest and the whole mass surges onto the foreshore. On shallower slopes, the crest may spill into the trough in front, or—creating the "tube" sought by surfers—rear up into a crest as the trough retreats, crashing down, trapping air beneath it, and hurtling up the beach. Waves conspire with tides to create currents along the shore or, if the space is constrained, rip currents that move rapidly out to sea, the only direction in which the mass of water can go.

The shape and character of a beach will typically change dramatically between winter and summer. In summer, the foreshore slope is gentler, the waves are less steep, and net sand movement is onshore. In winter, more violent wave action steepens the slope and the beach is eroded, with the sand being transported offshore to accumulate in offshore bars. Winter storms will throw up a steep bank of coarse-grained debris at the back of the beach, the *berm,* which remains beyond the reach of waves during the summer. The mapmakers really should start all over again at least twice a year.

Beaches come and go naturally, either seasonally or over longer periods. The gloriously named Five-Finger Strand in Donegal, northwestern Ireland, has long been a local attraction but a frustratingly fleeting one—every once in a while it disappears for a few years, leaving an unattractive expanse of pebbles. Its most recent disappearance was, predictably, marked by local demands for action, but, fortunately, researchers at the University of Ulster were able to scrutinize the ledgers of the sediment budget for the area. They found that the stability of the beach depends on the precise position of a local inlet—when the inlet shifts, there's a budgetary demand for sand and the beach supplies it. After a few more years, the inlet shifts again, the demand is no longer there, and the offshore accumulation disintegrates, returning sand to the beach. This particular geological prediction—that the sand would return to Five-Finger Strand—was positive and correct. Five-Finger Strand was not an urgent case—no human infrastructure was threatened, just leisure—and it was nature's own infrastructure that was the problem.

The approach used by the researchers investigating Five-Finger Strand reflects the definition of what are termed *littoral cells,* compartments of a coast where the sand budget can be defined in terms of three elements—sources, transport, and deposition—and is more or less isolated from what is going on in neighboring cells. Sand eroded from a beach, a headland, or a pile of glacial debris does not simply travel along the coast endlessly—it may be taken out of circulation for a while by being dumped into a washover fan or swept out onto the shelf and, as we shall see, the deep ocean. Shoreline topography, waves, and currents define the extent of a particular littoral cell, which may extend for a hundred kilometers along the coast and contain its own subcells. This concept has proved of particular value in understanding complex coasts such as that of California—and a number of surprising discoveries have highlighted how much we still have to learn.

For example, it had always been assumed that the load carried by California's rivers was the primary source of sand for its fabled beaches. But researchers at the University of California at San Diego looked carefully at the littoral cell that covers 80 kilometers (50 mi) of coast in the La Jolla area. Using forensic sand-grain studies, combined with detailed imaging of the sea floor, they showed that at least half the beach sand is derived from the erosion of coastal cliffs and promontories. The rivers may periodically flood huge amounts of sand to the coast, but only when it rains—and not only is that an increasingly infrequent occurrence, but dams, desert irrigation, and concrete channels have also profoundly modified natural sediment delivery. Coastal cliff erosion continues year in, year out, but since many of these critical sources are artificially "armored" to preserve their scenic attraction, beaches may be starved of sand supply as a result.

This is hardly a minor result and demonstrates how far from perfect our understanding of coastal sand movement really is. Nevertheless, we continue to plunge ahead with planned or unplanned amendments to sand budgets, creating a spiraling sequence of cause and effect. For example, travel northward from La Jolla to a different cell, the one that extends from San Francisco's Golden Gate Park down along the coast past Ocean Beach. Golden Gate Park was once an expanse of sand dunes, but it is no longer, and Ocean Beach, once naturally backed by dunes, is now lined with houses. San Francisco Bay is the natural conduit for huge amounts of sediment, but because the buildup interferes with navigation, the bay is dredged and the sediment is dumped offshore. The budget is completely busted. Ocean Beach dunes no longer provide a buffer against which waves can dissipate their energy, so instead the waves erode the beach. The supply of sand from the bay has been vastly reduced, and sediment movements have been changed by the artificial sandbar created by dumping the sediment dredged from the shipping channel. At least sand is no longer mined from the northern end of the beach, as it was (for gold) early in the last century. The solution would seem to be a budgetary enhancement via the creation of a different artificial sandbar, and the Army Corps of Engineers has been dumping several tens of thousands of dump truck loads of dredged sediment in a new location, closer to Ocean Beach. The results are "promising"—parts of the beach are growing, but is this the result of the new sandbar? Only one thing is certain: it's a whole new budget, and it will be years before we know if it balances or not.

NOURISHMENT

Occasionally, human activities will change the budget in a constructive way. Crinnis Beach, a popular tourist spot in Carlyon Bay, Cornwall, did not exist before the nineteenth century. Mining the china clay formed from the rotting granites inland had been taking place for hundreds of years, and the sedimentary debris from these mines, carried down to the coast by rivers, threatened harbors and estuaries. In the mid-nineteenth century, one of the rivers was diverted into Carlyon Bay, creating a beach where there had not been one before. Now that mining has ceased, the natural budget requires that the beach be removed, much to the concern of the local tourism industry, and remedial efforts are underway.

Beach nourishment is the official term for trying to put the sand back artificially on shrinking beaches. Sand in gigantic quantities is dredged or pumped via monster hoses from offshore and sprayed onto the eroded beach, at a cost of up to $10 million per mile. On the U.S. east coast barrier island chain, from the southern end of Long Island to southern Florida, the equivalent of 23 million dump trucks of sand has been spread over 195 beaches on more than seven hundred separate occasions. Virginia Beach has been "renourished" more than fifty times. A fundamental problem is not simply that this is treating the symptoms rather than effecting a cure, but that the wrong kind of medicine is often used. Beach sand is, after all, the specific local product of local processes, and replacing it with sediment dredged from offshore is replacing it with something else entirely—finer-grained sand, differently sorted sand, or mud. A typical "nourished" beach erodes much faster than a natural one. There are examples where the best sand has correctly been identified as that in tidal inlets—after all, it's sand that derives from longshore drift along the beach. But remove that sand, and the only thing that is created is another problem—the next beach downstream from the inlet has a reduced supply.

This is a problem the world over. After the beaches in Cancún, Mexico, were removed (again) by Hurricane Wilma in 2005, tourism was devastated. Renourishment cost $19 million, but a year later the beaches had shrunk to less than 20 meters (65 ft) and the waves had cut steep ledges, which detracted significantly from the beach experience. Sandbags, ironically, are now in place. The hotels suffer, but they are a cause of the depletion, having replaced the vegetation that naturally helped hold the sand in place. Hurricanes always have, and always will, remove beaches; waves always have, and always will, erode the coast. Sea level is rising, and much of the world's coastline is eroding.

As a very small child, I enjoyed family holidays on the beach at Sheringham, a resort on the east coast of England. Engaged in digging holes and building sand castles, I was blissfully unaware of the erosional conflict going on around me. I was equally unaware of the descriptions of Sheringham by one of the nineteenth-century founders of modern geology, Charles Lyell. In 1830, Lyell published *Prin-*

FIGURE 25. Aerial photographs of coastal erosion in eastern England. The arrows mark the same point; note the dramatic recession of the cliffs. (Photos courtesy of Environment Agency of England and Wales and the Coastal Concern Action Group)

ciples of Geology; or, The Modern Changes of the Earth and Its Inhabitants Considered as Illustrative of Geology. This is arguably the most influential book in the history of geology as a science—it ran to eleven editions during Lyell's lifetime, each substantially extended with new material. The book was carried by Charles Darwin on the *Beagle* voyages. Indeed, Lyell did for our understanding of the Earth what Darwin was to do for biology, and he certainly understood coastal change:

> The waves constantly undermine the low chalk cliffs, covered with sand and clay, between Weybourne and Sherringham [sic], a certain portion of them being annually removed. . . . Between the years 1824 and 1829, no less than seventeen yards were swept away, and only a small garden was then left between the building and the sea. There

was, in 1829, a depth of twenty feet (sufficient to float a frigate) at one point in the harbour of that port, where, only forty-eight years before, there stood a cliff fifty feet high, with houses upon it!

I was digging in the sand before the catastrophic storms of 1953, which claimed over two thousand lives and caused what is benignly referred to as *coastal realign-ment*. It continues, day in, day out. Figure 25 shows another set of "before" and "after" aerial photographs, this time of the hapless village of Happisburgh, just up the coast from Sheringham. But they were not taken "before" and "after" a single destruc-tive event; they are simply two frames in the film of ongoing, destructively rapid coastal change. Look at the shore in the two photographs: groins have been built to protect the coast, but they have simply disrupted the natural flow of sand, and the sea has destroyed many of them. "Coastal defenses" simply reorganize the lit-toral cell and the sediment budget—the problem is either exacerbated, modified, or shifted elsewhere. The sea takes *its* nourishment from wherever it wishes.

Natural coasts do not need defenses or protection; they change as the rules of the game and the endless dance require. Coastal change has become a problem only since we elected to ignore the prudence of earlier times; civilizations developed along coasts, but with caution. Constrained between the wilderness, the wrathful gods and monsters of the mountains behind, and the visible power of the sea, our ancestors chose to live in the great in-between. We now choose to ignore that experience— and to feel that somehow or other we can do something about the consequences.

BEACH DETAILS

Coastal dynamics are the large-scale movements of the dance, but for a holiday visitor, it's the details that are most obvious. Any beach is a canvas of patterns and shapes, the swash and backwash of the waves, dark mineral brushstrokes, rivulets, bubbles, pits, and footprints.

Often a beach is not flat, but covered in small ripples, miniature waves in the sand (Figure 26). They may be simple long parallel ridges or sinuous oval shapes. These are bedforms in the sand that result from the interaction between the fluid moving over the grains and the grains themselves—and, yet again, their origin is

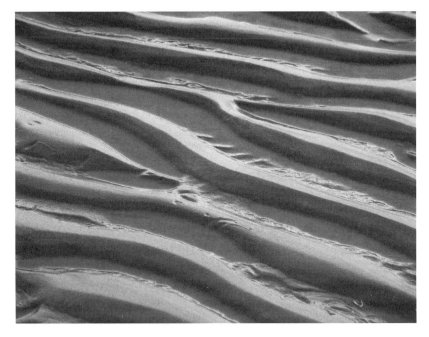

FIGURE 26. Beach ripples. (Photo by author)

a matter of some debate. You would have seen similar ripples on the surface of point bars in the Susquehanna, formed by the one-way flow of the river—current ripples, asymmetrical, with steep sides in the direction of current flow. But the ripples on the beach are different— generally much more symmetrical, and often with sharp ridges. In the simplest sense, a wave moving back and forth over the sand transfers its energy to the grains, washes them to and fro, building up small ridges, across which the grains move with the waves. But it's not that straightforward. For a start, many stretches of beach where the waves move back and forth are flat and unrippled, and why ripples develop in some circumstances and not in others is not entirely clear.

George Darwin, son of Charles, was fascinated by ripples and performed a series of experiments using ink in water to define the patterns of turbulence and eddies that move the sand grains. Ralph Bagnold made detailed experiments that reproduced symmetrical ripples and developed relationships between fluid dynamics, ripple size, and grain size. Internally, all ripples show exotic variations of cross-bedding on a small scale, providing clues when such structures are found in ancient rocks. We can describe and name the variety of ripple shapes; we can see that some complex forms result from the combined action of waves and currents. But we don't really understand them. Larger wavelike forms can develop—water-formed dunes, often with ripples superimposed on their surfaces.

Dark streaks of black sand often serve to emphasize ripples on the beach. As sand grains are repeatedly picked up and dropped by the waves, the heavier grains respond differently than the lighter ones. Grains of iron minerals, heavy for their size, will be the first to be dropped as the wave's energy diminishes. And so these dark smears are clusters of heavy minerals, and they often are iron—as observed in the eighteenth century by Petrus van Muschenbroek. If the winnowing of the heavy grains is allowed to continue, then over time concentrated deposits of these minerals will form. This has happened in the geological past, and commercially exploitable mineral deposits may result, as we shall see in chapter 9. This also occasionally happens on a modern beach, and such an occurrence was the cause of one of Thomas Edison's many business failures. On a fishing trip with friends off the coast of Long Island, Edison put into shore for lunch and found the beach covered with a layer of black sand. He took some home and discovered that the black grains were a magnetic iron oxide mineral—magnetite—which stuck to a magnet while the common sand grains fell off. Edison's enthusiasm ran, as it often did, ahead of his business sense, and he immediately arranged for the purchase of the beach and the manufacture of separating machinery. Unfortunately, by the time he and his colleagues returned to Long Island, a winter storm had reworked the beach and completely removed the black sand.

Every wave breaking on the beach acts as a pump, forcing water through the sand and expelling the air between the grains; when the wave recedes, the water drains out, to be replaced once more with air. This constant intergranular violence makes the tenacity and inventiveness of the meiofaunal residents even more remarkable. Commonly on the beach, you will see large numbers of small holes and some miniature volcanic pits and craters; some holes are made by the residents, but many are the result of the incessant pumping and flushing action of the waves and mark the expulsion of air from the sand. Waves breaking can trap air, making the pumping action more complex and, together with stirred-up organic products, forming the bubbles of sandy foam that sometimes make the beach look like a drained bubble bath. Waves breaking at an angle to the beach will drain back straight down the slope, leaving behind overlapping curving patterns of heavy minerals and debris. A beach is a place of a compelling myriad of patterns and processes, not all of them understood.

RESIDENTS AND GUESTS

Many of the marks and patterns of a beach are, of course, made by the residents— or visitors. Large pits in a beach in Washington State were found to have been left by gray whales excavating the sand at high tide to filter out ghost shrimp buried among the grains; the sand presumably was expelled elsewhere—an unusual means of sediment transport. The shrimp are among several kinds of "ghosts" that may inhabit a beach, rarely seen but clearly evident. Ghost shrimp burrow in the sand below the mid-tide zone to a depth of 2 meters (6 ft). Their burrows, surrounded by small piles of excavated and excreted sand, are evident when the tide goes out. Ghost crabs make complex multichambered residences below the beach or dunes, the entrance hole quite large and surrounded by balls and small piles of sand. Ghost crabs seem to love playing with the sand that they have stripped of food, apparently simply for the fun of it, tossing it about (Plate 8), forming and hurling small balls of it or arranging them into rows. When the crabs emerge from their burrows, their camouflage makes them difficult to see, as does their pace: on hard-packed sand, ghost crabs can move at up to 2 meters (6 ft) per second, a sudden blur of movement. On softer sand, however, they feel the effects of the character of granular materials and slow down markedly.

Holes in the beach can be made by shrimp, crabs, worms, doodlebugs, beetles, snails, shellfish, and a huge variety of other creatures that make their homes among the grains, excavating, building, filtering, and excreting. The microscopic meiofauna of a healthy beach that we saw in chapter 2 provide a richly stocked larder and represent the small-scale end of a long chain of life that tends the beach. Many of these creatures use sand grains to construct their residences, but perhaps the most extraordinary is the trumpet worm, a member of the class Polychaeta, or "many bristles." The bristles are remarkably strong yet delicate sand-grain movers, both bulldozers and food cleaners, but they are also building tools for the trumpet worm, a master craftsman. From an early age, the worm constructs a protective conical (trumpet-shaped) shell around itself, grain by grain. Each grain is individually selected, examined, and discarded or fitted in place. The process is meticulous, each grain being rotated or repositioned to accomplish a perfect fit, and only then is the glue, secreted from a gland near the worm's mouth, applied. The structures can be several centimeters (2 to 3 in) long (Plate 8). After the worm dies, the shell can sometimes be found on the beach before it is recycled by the waves.

Not only do the creatures that live under the sand find a ready food supply in their cohabitors, the meiofauna, but their subsurface living also allows them to avoid aboveground predators. The beach is a larder for birds, as described so lyrically by Elizabeth Bishop in her poem "Sandpiper":

> The roaring alongside he takes for granted,
> and that every so often the world is bound to shake.
> He runs, he runs to the south, finical, awkward,
> in a state of controlled panic, a student of Blake.
>
> The beach hisses like fat. On his left, a sheet
> of interrupting water comes and goes
> and glazes over his dark and brittle feet.
> He runs, he runs straight through it, watching his toes.
>
> —Watching, rather, the spaces of sand between them,
> where (no detail too small) the Atlantic drains
> rapidly backwards and downwards. As he runs,
> he stares at the dragging grains.
>
> The world is a mist. And then the world is
> minute and vast and clear. The tide
> is higher or lower. He couldn't tell you which.
> His beak is focussed; he is preoccupied,
>
> looking for something, something, something.
> Poor bird, he is obsessed!
> The millions of grains are black, white, tan, and gray,
> mixed with quartz grains, rose and amethyst.

SAND SEAS BY THE SEA: DUNES

As at Kitty Hawk and, once upon a time, at California's Ocean Beach, great ridges of sand dunes often form the backdrop to a beach. Sand is, after all, plentiful, and the wind blows constantly along the coast, reworking the beach into hills of sand. Coastal dunes form a barrier along the coast, buffering the land or water behind by absorbing the energy of storm waves and breaking the force of the wind. They are very different from desert dunes, often less well organized and subject to different forces from different directions. But they are just as mobile as their arid relatives and can be nearly as massive: dunes along Lake Michigan are more than 100 meters (330 ft) high, and those along the Oregon coast extend up to 5 kilometers (3 mi) inland.

Because of their mobility, coastal dune systems are often described as "fragile" environments, but this perspective arises only from the palpable impact of their mobility on human structures—it's the human structures that are fragile, not the dunes themselves, which are designed by nature to change constantly. Because of the intimate proximity of coastal dunes to human activity and development, conflict is more common than it is with desert dunes, and more dramatic, as noted by Robert Frost in his "Sand Dunes":

> Sea waves are green and wet,
> But up from where they die,
> Rise others vaster yet,
> And those are brown and dry.
>
> They are the sea made land
> To come at the fisher town,
> And bury in solid sand
> The men she could not drown.

Yet human activity can also create coastal dunes: today's spectacular dunes at Cape Cod were far less spectacular at the time that the first Europeans arrived, but forest clearing and grazing cattle destroyed soil stability, liberating sand to build the dunes.

We shall look in some detail at the work of the wind and how sand dunes behave in the next chapter, but the principles are simple and follow many of the same rules as for the transport of sand in any medium. As a wind gathers force across a drying beach, it develops sufficient energy to roll, bounce, or pick up sand grains; as the wind loses energy or as the sand grains embed themselves in an already-existing dune, the sand is deposited. But unlike in the desert, water plays a role in the evolution of the dunes. Storm waves erode them, and even in calm conditions rain falls on them, conspiring with salt to form hardened layers that can resist erosion. Vegetation specializing in sandy and salty conditions can capitalize on periods of dune stability and fix the sand even more permanently in place. Many active dune ridges along the beach are backed by vegetated and stabilized ancestors. In some situations, the dunes advance seaward, overriding the beach; in others—barrier islands, for example—they are part of the mass migration landward. On the Oregon coast, forests have been buried as the dunes migrate southward.

Whereas Stephen "Dr. Beach" Leatherman rated Ocracoke, in 2007, as the country's most beautiful beach, and different stretches of the Brazilian coast have been described as "the world's best beach," Daytona Beach, Florida, bills itself as "the world's most famous beach." In the 1930s, Daytona Beach was renowned as the setting for a series of land speed records. Wide, sweeping beaches went hand in hand with the extensive dune ramparts that extended the length of Florida's island coast around Daytona, the two systems existing in a kind of sandy symbiosis. Today, those dunes that have not been bulldozed or built over have largely been eroded as a result of sediment budgetary changes. Where once land speed records were set on a wide beach, the coastal highway is now regularly taken out by storm erosion.

On the opposite coast from Daytona, erosion of the sand dunes protecting the property of the celebrity residents of Malibu Beach has prompted them to bulldoze sand from the public beach to re-create artificial "dunes," an action greeted, not surprisingly, by considerable public uproar. Elsewhere in the world, moving sand, even with the best of intentions, is also frowned on. The coast of northwestern England, north of Blackpool, is a great stretch of broad beaches and dunes, constantly worked by the stiff breezes of the English seaside. Arthur Bulmer, a pensioner, had enjoyed his modest beachfront home until a week of storms blew seven tons of sand into his garden. The same storms had moved huge quantities of sand into town, building up dunes in car parks and covering local roads. While the local authorities are responsible for clearing the roads, private gardens are the responsibility of the owner;

the obvious thing to do, thought Mr. Bulmer, would be to take the sand back to the beach, wheelbarrow by wheelbarrow. This would seem to constitute a fulfillment of civic duty and demonstrate responsibility for environmental restoration. But he was told by the authorities that this would be illegal; it would be an act of *fly-tipping*, the quaint British term for illegal dumping of refuse. Mr. Bulmer would be liable for the equivalent of a $100,000 fine and up to six months in prison—and have his vehicle, the wheelbarrow, confiscated. His only option was to hire a professional waste disposal company to remove the sand to a waste disposal site.

Arthur Bulmer's stretch of coast is also known as "England's Golf Coast," illustrating a further benefit of preserving back-beach dune ridges. The Old English word for ridges is *hlincas,* from which the term *links* is derived, and it was in this terrain that small leather balls stuffed with goose feathers were first whacked about and the revered game of golf—or *gouff,* or *goffe*—originated. Many of the world's finest golf courses, including the Old Course at the Royal and Ancient Golf Club at St. Andrews on the east coast of Scotland, owe their location to coastal dune systems. But nature plays her own game: the seventh and eighth holes of England's oldest golf course, set in the dunes of the north Devon coast, are about to disappear beneath the waves. On the west coast of Scotland, Alfred Nobel used the coastal dunes around Ardeer as the natural location to construct not golf bunkers, but explosives bunkers, locating his dynamite manufacturing facility there in 1873.

The sheer vastness and beauty of some coastal dune systems is extraordinary. The exquisite photographs of one of the twentieth century's most influential photographers, Edward Weston, and those of his son Brett celebrate the contours and sculpture of the dunes at Oceano in Southern California. The forms and shadows of the sand are haunting, the grains often seeming to be individually visible, the constant movement caught on film. And the sensual forms of sand are echoed in the nude studies. Unfortunately, a different kind of resonance is now more common: the dunes at Oceano are open today to four-wheel-drive and off-road vehicle recreation.

THREATS AND PLEASURES

Coastal dune systems, because of their close relationships with human occupation, provide compelling images of sand as a *threat.* On the south coast of Wales, the

village of Kenfig was established in the twelfth century as a small port and farming community on a river sheltered by the coastal dunes. It survived the attacks of
marauding Welsh tribes, but not the forces of nature, provoked by grazing and the
ensuing destabilization of the sand. These factors, probably combined with climate
change as the cold period often referred to as "the Little Ice Age" approached, meant
that the dunes were on the move. By the fourteenth century, large parts of the
town and its fields were covered, and by the middle of the seventeenth century, it
was completely abandoned. Hans Christian Andersen's tragic story *Skagen* is a tale
of death, evil-doing, and unrequited love among settlements and farms overrun
by shifting sands (Skagen church remains today half-buried). The story was sung
to him "by the storm among the sand dunes." As Henry David Thoreau wrote in
Cape Cod, "The sand is the great enemy here."

Sand as the enemy is the theme of two stunning, surreal, award-winning films.
Woman in the Dunes, the 1964 film by Hiroshi Teshigahara, won the Grand Jury
Prize at Cannes and was nominated for two Oscars. Based on the novel by Kobo
Abe, it is set in the sweeping wilderness of sand dunes along the Japanese coast at
Tottori. An amateur insect collector finds himself abducted and consigned to co-
habit with a woman whose crumbling house lies at the bottom of a deep depression in the dunes; she exists "only for the purpose of clearing away the sand." The
sand constantly threatens the house, an outpost on the edge of a village "already
corroded by the sand," whose way of life is defined by fighting the dunes and selling the salt-tainted sand—illegally—to construction companies. In both the book
and the film, sand is the enemy but also the central character. The meticulous cinematography and Abe's writing beautifully document sand grains and their movement: liquid avalanches descending on the house, windblown grains, and the forms
of shifting dunes. The book is a lesson in the behaviors of granular materials. And,
inevitably, the sand is a powerful metaphor for time: "Monotonous weeks of sand
and night had gone by."

Today, the dunes at Tottori are shrinking due to withdrawals from the local sediment budget, but the 1,000 square kilometers (400 sq mi) of glistening white dunes
of Lençóis Maranhenses National Park on the northeast coast of Brazil appear under no such threat. A photograph hanging in a bar on the dune-strewn coast of
nearby Ceará was the inspiration for Andrucha Waddington's *The House of Sand,*

a prize winner at the 2006 Sundance Film Festival. The photograph showed a dilapidated house, once owned by a Brazilian "woman in the dunes" who had spent her life fighting the sand. By the time she died, the house had been destroyed; the sand had won. The film, shot entirely in Lençóis Maranhenses National Park, has four main roles: three generations of women, and the sand. As in Teshigahara's film, the dunes provide both sweeping landscapes for stunning cinematography and constant microactivity: trickling, avalanching, and flying grains. *The House of Sand* is a spectacular and haunting illustration of the use of sand as a metaphor for time, isolation, emptiness, and change.

But sand can also be a friend. People love beaches and even create them in the most unnatural places. In 1934, 1,500 tons of sand were dumped onto the mudflats of the Thames below the Tower of London. Intended to provide the only taste of the seaside that many of the city's poor would experience, Tower Beach was a great success. Even though it could accommodate only around five hundred people and was useable only at low tide, hundreds of thousands of people flocked to it over the following five years, building sand castles, paddling in the water (of dubious quality), relaxing in deck chairs, and enjoying typical beach entertainment. The beach was reopened after the war, but it finally closed in 1971.

The sand below the Tower of London was arguably the first *urbeach*—a recently coined term to describe beaches constructed in the middle of cities, an increasingly popular habit. Paris has created the Paris Plage along the banks of the Seine every summer since 2002; Berlin, Amsterdam, Rome, New York, and Las Vegas (complete with waves) have followed suit. In the United Kingdom, the most landlocked and urban of cities, Birmingham, is spending $400,000 on two urbeaches, and an indoor beach was a 2007 summer feature of the newly refurbished Millennium Dome.

OUT TO SEA

Armies of sand grains move up and down the coast. They may be trapped en route for a season or for eons, but many are swept by storms, tides, and currents out onto the shelf—and beyond. Tides are not simply the sea rising and falling, but are themselves huge, slowly moving waves. Around the British Isles, this tidal pulse moves clockwise, up around Scotland and down the North Sea. Coastal and seabed topography conspires with this movement to create strong local tidal currents, traveling at speeds of a meter per second (2 mi/hr) or greater, more than sufficient to move huge quantities of sediment. All this complexity and large-scale sloshing around leads to places on the seabed where the net water movement diverges, a "bed load parting." A sand grain blown into the Seine from the Paris Plage and flushed out into the English Channel (or, since it is a French sand grain, La Manche) could find itself, depending on the time of day and the time of year, traveling out into the Atlantic or up into the North Sea.

Storms and the surges of water that often accompany them supply sand from the coast to shallow marine waters and the engines of tidal currents. The results are vast offshore landscapes of sand: ridges, ribbons, underwater dunes (or sand waves), and bars, constantly on the move. Fields of huge sand waves have recently been discovered covering the floor of the outer reaches of San Francisco Bay; in 100 meters (300 ft) of water, individual wave forms are up to 200 meters (700 ft) long and 10 meters (30 ft) high, moving at 7 meters (23 ft) a year. Huge volumes of sediment (much of which probably originated from human erosional activity during the Gold Rush) have been flushed through the bay, building the field of sand waves. Modern reduction of this supply of sand undoubtedly contributes to the problems we have seen at Ocean Beach. Sand waves are typical of areas where tidal pulses are strong: those in Puget Sound, Washington, are three times the size of the San Francisco examples.

In shallow waters, these bedforms constitute major shipping hazards at high tide but may be exposed at low water. One of the world's iconic paintings, Théodore Géricault's *The Raft of the Medusa* (1818–19), portrays the desperation of survivors of a real event, a shipwreck on a sandbar off the coast of Senegal in 1816. The infamous Goodwin Sands, 16 kilometers (10 mi) of sandbank off the southeast coast of England, have claimed perhaps two thousand ships, been cited as the site of lost communities, are inhabited by countless ghosts and other spiritual manifestations, and, until recently, were the ground for an annual cricket match. The strange English tradition of playing cricket in the sea persists today farther west: sailing for the first time recently into the Solent, the great estuary of Southampton, a Dutch sailor ran aground on Brambles Bank. Seeing a small flotilla approaching, he took them to be intent on his rescue, but instead the passengers disembarked and set up the stumps and a scoreboard for a game of cricket. Such sporting events are, of course, cut short by the tidal pulse.

The continental shelves, and particularly those fed with the seemingly endless supply of glacial debris, are temporary resting places for the majority of marine sands. Sample the seabed anywhere around the British Isles and the chances of bringing up anything other than sand of some sort are small. The same is true of large areas of all the Atlantic continental shelves. Stretching from the Grand Banks of Newfoundland to the shores of New England is a series of massive sandbanks, constantly shifting across the seabed between the coast and the edge of the continental shelf, the ongoing resculpting of glacial debris. Uniquely, and for reasons not entirely clear, these sandbanks emerge above sea level to form Sable Island, 160 kilometers (100 mi) southeast of the nearest mainland. This is one of the world's most extraordinary—and dangerous—islands. Positioned in the open North Atlantic, subjected to gales, hurricanes, pounding waves, and ferocious winds, and constructed of loose sand without any solid foundation, it seems that it has hardly any right to survive. But survive it does, and it's a dynamic place.

Sable Island is long and thin, around 40 kilometers (25 mi) from tip to tip (although measurements vary) and just over a kilometer (little more than half a mile) across at its widest point. It lies roughly east-west, its crescent shape looking like a thin-lipped smile in the middle of the ocean. But the smile is deceiving: the island is simply the tip of a vast iceberg of sand, the much, much larger Sable Island Bank. Just beneath the ocean's surface is a rolling topography of moving sand stretching up to 14 kilometers (9 mi) north to south and perhaps 30 kilometers (19 mi) east to west. The island lies at the meeting point of three major ocean currents, the Labrador current moving southward, the St. Lawrence current moving eastward, and the Gulf Stream on its way north. This interaction causes the waters of the Atlantic to swirl in a great counterclockwise pirouette around the island; but bottom currents, controlled by the tides, swirl clockwise. The result is a complex of ridges and valleys of sand running roughly northeast to southwest. This is a shape-shifting topography, the ridges migrating 50 meters (160 ft) every year, and changing rapidly during the frequent storms. Nautical charts are barely worth the paper they are printed on.

Is the whole island on the move? Opinions differ, and seasonal and longer-term changes in its length and width may create an illusion of wholesale migration. It is located close to the edge of the continental shelf and just south of the Gully, an immense submarine canyon carved down through the edge of the shelf when sea level was low. It seems possible that, over time, large areas of the bank, and perhaps Sable Island itself, will end up as a deluge of sand over the side of the canyon to be swept out onto the deep ocean floor (more on that later).

For the moment, though, the vast shallow sandbanks that surround Sable Island protect it from destruction, breaking the force of the waves in all but the most violent storms. It survives but changes constantly with the seasons and the weather. A hundred years ago, there was a long lagoon sheltered within the sand dunes of the island, but today storm waves have broken through the barrier and the lagoon is essentially filled up with sand. Remarkably, the island's sand does trap freshwater from the rains of the North Atlantic; cattle were once raised on the island, and a population of feral horses has survived since 1738, eking out a living on the sparse vegetation of the dunes.

The island also has a dramatic human history. Its name is the French word for "sand," and earlier it was called Isola della Rena, *rena* being Italian for "sand." Situated in prolific fishing grounds and on the great circle route from Europe to North America, it has been encountered, not always happily, by generations of sailors, perhaps since the Vikings. It is one of a number of places that claim to be "the graveyard of the Atlantic," but the island's claim is well founded. Wrecks, hundreds of them, tragically chart much of the human history of the island. The fogs of the northwestern Atlantic are notorious, as are the storms. Ships have been driven onto the Sable Island Bank or simply sailed straight onto its ever-shifting topography. Some have suggested that the sands contain enough grains of magnetic iron minerals to throw off ships' compasses; there are indeed magnetic grains concentrated in parts of the island and its surrounding sandbanks, but not enough.

Perhaps the most infamous event with which Sable Island is associated took place in October 1991. Three major storm systems, one a hurricane off the east coast of the United States, the others major cold weather depressions in the northwestern Atlantic, joined forces to create one gigantic storm. Gathering force quickly, this storm arose too quickly to be officially named, but it has since been known as "the Halloween Storm," or, more widely recognized, "the Perfect Storm." Record waves, tides, and winds resulted up and down the coast from this meteorological conspiracy, as did a book and a film. The tragic story of the *Andrea Gail,* a swordfishing boat out of Gloucester, Massachusetts, became the stuff of big-screen drama. The only wreckage traceable to the ship, small pieces of equipment, was found washed up on the shores of Sable Island.

THE TROPICS

So far, reflecting their geographic dominance, we have considered the coasts and shelves of higher latitudes, ruled by quartz-rich sand and the legacy of glaciation. Tropical and subtropical systems play, in most ways, by similar rules—after all, a sand grain is a matter of size, not what it is made of, and transport by water or wind will be essentially the same whether it is an old quartz crystal or a broken piece of shell. Tropical sands form beaches, dunes, barrier islands, washover fans, and tidal bars in exactly the same way that their more northerly relatives do. But there are some differences. Tropical seas are dominated by marine life, much of which makes hard parts—shells and skeletons—out of calcium carbonate, and it is these, together with fish feces, that provide much of the raw material for the local sand. Calcium carbonate is a very different mineral from quartz—less hardy, more soluble, and more vulnerable to chemical processes. But some aspects of its behavior are odd. Most soluble substances—salt, for example—dissolve more easily the higher the temperature of the water, but calcium carbonate does the opposite: the warmer the water, the less soluble it is, which is why water-heating appliances become clogged with carbonate scale. It is also more soluble in water that already contains dissolved carbon dioxide; hence it is more soluble in seawater than fresh. These properties profoundly affect the ways in which tropical marine creatures regulate their shell construction, what happens to the shell as it sinks into colder water after the occupant dies, and the chemical budget for carbonate sediments. Tropical marine environments are carbonate factories, manufacturing, distributing, and consuming it.

Virtually all of the sand in tropical environments is locally produced; the northerly characteristics of long-distance transport by rivers and along coasts are far less important. In chapter 1, we saw some examples of sands from tropical environments, including foraminifera skeletons, coral fragments, and ooliths. Ooliths, formed by the chemical precipitation of calcium carbonate growing around small grains that are rolled gently in the balmy tropical waters, assemble in huge numbers to form some of the major marine landscapes of the tropics (see Plate 4).

One of the most studied carbonate factories is that surrounding Florida and the Bahamas (the environment for field investigation being particularly attractive). This region may have been far from the direct devastation of the glaciers and their aftermath, but it was not immune to the lowering of sea level. In Plate 7, the vast extent of the shallow water shelf (in light blue) is clear, and during the ice ages, it would have been largely above sea level. Myriad coral reefs were rudely exposed, then died and disintegrated, providing their own version of glacial debris. As sea level rose, this was the grist for the tropical mill. The Florida Keys (or "cays," from *cayos*, Spanish for "small islands") are made of the modern corals that reestablished themselves with the renewed sea level, surrounded by their own modern coral debris as well as that of their ancestors, combined with the ancient sands from the glacial periods, shell fragments, and ooliths. Together, the Keys and the Bahamas comprise hundreds of these islands, and the surrounding waters are treacherous—another "graveyard of the Atlantic."

During their exposure to the atmosphere, many carbonate landscapes—reefs, dunes, and shoals—were *lithified,* transformed into solid rock, the grains often cemented by carbonate. The resulting limestone commonly forms the foundation of the modern islands of the Bahama Banks. But the Banks have a much longer history, having been building up carbonate sands, reefs, and other sediments for tens or perhaps hundreds of millions of years. The region subsided under the sheer weight of this enormous pile of sediments, and it is constantly making room for further accumulation, producing a giant carbonate platform building out into the Atlantic. The total thickness of this accumulation beneath the Bahamas is 5 kilometers (16,000 ft) or more, over an area of 132,000 square kilometers (more than 50,000 sq mi). And in addition to providing vacation delights, this great carbonate factory also does the planet a favor: it extracts large volumes of carbon dioxide from the atmosphere.

INTO THE DEEP OCEAN

When HMS *Challenger* embarked on the maiden voyage of modern oceanography, it was believed not only that the deep ocean floors were lifeless, but also that they were sandless, since scientists could think of no conceivable means of moving sand grains out into the deep sea. The researchers aboard the *Challenger* disproved this by collecting occasional sand samples from the deep ocean floor. Their sampling was hardly comprehensive: the grains they found may have been dropped from melting icebergs, one obvious means of getting sand out there, but they may have accidentally encountered more extensive deposits of sand. In the following decades, thanks to serendipitous events, it was realized that there are routine paths for large volumes of sand to follow off the shelf, down the slope, and out into the deep.

As noted earlier, one of the many reasons for the *Challenger*'s voyage was to provide data that would facilitate the stringing of telegraph cables across the world's oceans. In 1858, fourteen years before the departure of the *Challenger,* England and America were first linked, albeit briefly, by wire. Two ships joined their supply of wire in the middle of the Atlantic and then sailed off in opposite directions. The connection enabled Queen Victoria to send a message to President Buchanan, a nineteen-hour transmission but still faster than a letter. The line failed soon after, and it was not until 1868 that the first commercial link was made, between the United Kingdom and Canada. Meanwhile, Britain was busily wiring itself up to its empire, and a host of cables from India and the Far East came ashore at the tiny sandy Cornish bay of Porthcurno. This location was chosen rather than the busy neighboring port of Falmouth because the cables could be securely buried in the sand away from the dangers of ships and their anchors. The Falmouth, Gibraltar and Malta Cable Company, which established itself there, would later become the Cable and Wireless Company. But what does all this have to do with the journeys of sand to the deep sea?

By the 1920s, there were more than a dozen transatlantic cables, which, following the shortest route, passed along the base of the slope below the shelf of the Grand Banks of Newfoundland. But in 1929, disaster struck. A powerful earthquake hit the region, devastating the Newfoundland coast and rupturing twelve submarine cables. The exact timing of the ruptures was, of course, accurately recorded. It was thought that the earthquake was directly to blame, but it was not until twenty years later that Bruce Heezen, oceanographer and sea floor cartographer, and his colleague Maurice Ewing suggested the reason for some extraordinary aspects of the events. In Heezen's words, quoted by Arthur Holmes in his *Principles of Physical Geology* (1965):

> A study of the timetable of the breaks discloses a remarkable fact. While the cables lying within 60 miles of the epicenter of the quake broke instantly, farther away the breaks came in a delayed sequence. For more than 13 hours after the earthquake, cables farther and farther to the south of the epicenter went on breaking one by one in regular succession. . . . It seems quite clear that this series of events must indicate a submarine flow: the quake set in motion a gigantic avalanche of sediment on the steep continental slope, which broke the cables one after another as it rushed downslope and flowed onto the abyssal plain.

Since the positions of the cables and the timing of their breaks were accurately known, the speed of the avalanche could be calculated: up to 80 kilometers (50 mi) per hour. This was the first dramatic evidence for gravity, in this case aided by an earthquake, being responsible for the mass movement of sediment off the shelf and into the deep sea.

The cable breaks occurred at a deep trough in the seabed, over which the cables had been slung. The trough lines up with the Cabot Strait, the exit of the St. Lawrence into the Atlantic. This strait is not simply a modern body of water: it is also underlain by a huge valley, carved across the continental shelf by the ancestral St. Lawrence during the glacial periods of low sea level. That valley—the trough over which the cables were hanging—is incised into the slope and all the way out onto the deep-sea floor. And it is huge. Its sides are up to 1,200 meters (4,000 ft) high, not a lot less than those of the Grand Canyon. This submarine canyon carries sediment from the shelf to the deep, often very rapidly. It is only one of many canyons that cut the Atlantic continental slope: just to the south is the Gully, the route that the sand of Sable Island Bank follows to the deep sea. Many of these canyons are the extensions of river valleys carved over the shelf at the time of low sea level, themselves continuations of modern rivers like the Susquehanna.

The flows of sediment down these canyons are fast and turbulent, which makes them effective movers of material. After Heezen and Ewing's revolutionary study, a great deal of attention was paid to understanding ocean canyons and their behavior. Scientists from La Jolla documented the same features and phenomena off the coast of California, dramatically photographing waterfalls of sand cascading down even small canyons, ripples, and bed-forms associated with strong currents (see Shepard, 1973). The deepest submarine canyon in North America is California's Monterey Canyon, which *is* as deep as the Grand Canyon. As a significant part of the sediment budget for the local littoral cell, the Monterey Canyon siphons off an estimated dump truck's worth of sand every seventeen minutes—which is part of the reason that the shoreline erodes at an average of 15 to 30 centimeters (6 to 12 in) every year and monitoring equipment placed in the canyon doesn't last long.

Deep-sea drilling expeditions have sampled the sediments and imaged the canyons, demonstrating that in many ways they follow the same rules as river valleys, eroding, meandering, and flooding over their banks. The nature of the sediments deposited is, however, quite different in many ways. Water and sediment hurtle downslope in chaotic slurries, and the resulting deposits are the distinctive products of the complexity of gravity-driven turbid flow. Fine-grained sand can be transported many hundreds of kilometers out across the deep ocean floor by these *turbidity currents*. After the 1929 cable breaks, the sediment from the flow down the canyon covered an area of at least 100,000 square kilometers—the size of Kentucky or Iceland—and its volume and velocity caused a tsunami that was, rather than the earthquake itself, responsible for much of the damage in Newfoundland.

THE END OF THE ROAD

Down the Pacific coast, off the western shores of South America, things are very different. Even if a sand grain finds its way out across the narrow continental shelf of Central and South America, its travels come to an abrupt end. Here the slope is steep and seemingly endless, a plunge into the depths of the great trenches that line the edges of the continents, the places where the floor of the ocean is being consumed. Much of the Pacific is rimmed by trenches, and there is no way across them. Other than around the towering topography of the Hawaiian islands, there are virtually no deep-sea debris flows, and there is far less sand on the Pacific ocean floor than on the Atlantic. A sand grain tumbling into the trench may be scraped off and reincorporated into the continental crust as the oceanic curst is subducted, thrust beneath the continent. Or it will be carried down into the planet's interior, melted, and then, surging back toward the Earth's surface, it will eventually be re-born as a quartz crystal in a granite, waiting to be weathered and liberated. Sub-duction is the way in which sand grains truly die—but only to be reincarnated.

In the Atlantic, a sand grain (for example, our Susquehanna friend), if it has been flushed out across the shelf, along the ancestral river valley, and has cascaded down the slope, has not quite reached the end of the road. The great near-surface ocean currents, the Gulf Stream for example, are only part of the story of ocean circulation. Beneath these currents flows the other half of the "global conveyor"— deep ocean currents, driven by temperature and density contrasts, which follow the contours of the lower continental shelf. These deep currents, huge masses of moving water, can convey up to fifty times the flow of the world's rivers. They move at a sedate pace, but it is constant, and they are more than capable of reworking the sediment that has flowed out of the submarine canyons, carrying it along the deep edges of the continents, moving in the opposite direction from the surface currents. The deposits of fine sand that result, termed *contourites* because of their relationship with submarine topography, are now recognized as key records of the Earth's history.

But that really is the end of the road for our sand grain. Because of the great width of the ocean between the east coast of America and the submarine moun-tain range of the Mid-Atlantic Ridge, it will never reach the ridge; and even if it did, it could never find a way across it. To journey to Africa, our sand grain will have to wait for the planet to rearrange itself, for the Pacific to be consumed and so require the Atlantic to follow suit, until the sand grain, caught up in some fu-ture version of the Appalachians, is finally liberated once more by erosion and free to be carried eastward across whatever then constitutes the African continent. It will be a long wait.

SO WHY DOES ALL THIS MATTER?

The journeys of sand grains are stories of the way our planet works. Understanding and mapping those journeys is vital to the management of our immediate environment, providing lessons in what can and cannot—and what should and should not—be done. It allows us, as we shall see in chapter 7, to read the stories of our planet's history, to reconstruct the journeys of sand grains millions, even hundreds of millions, of years ago. But there is an even more fundamental moral to the story.

Philip H. Kuenen was an internationally renowned Dutch geologist. He was born in Scotland in 1902 but spent most of his life at the University of Groningen, developing research in marine geology. He made key contributions to the understanding of submarine canyons and their role in sediment transport and was a keen reader of the stories that sand tells us. In 1959, he was invited by the Geological Society of South Africa to give their annual lecture in memory of Alex du Toit, a South African geologist who gave early support to the ideas of continental drift. Kuenen chose as his lecture title "Sand—Its Origin, Transportation, Abrasion and Accumulation."

Kuenen addressed wide-ranging topics inspired by his subject matter, from the characteristics of an individual grain to its mass transport, illustrated throughout by startling, somewhat "back of the envelope" calculations on the number of new sand grains liberated per year, how long it takes to make a sand grain round, and so on. But he saved his most provocative thoughts for last. We have seen how, for sand, size is important, and that the size of sand grains depends on the typical size of the crystals in the granites and related rocks that are their parents. Kuenen took the implications of this a step further:

> The size of the sand grains determines the mode of transportation. If the particles were much bigger they could not be moved by the wind and there would be no coastal dunes or sand deserts. . . . The denudation of mountain chains would be a more laborious process. . . . The all-important part played by sand in protecting coasts from wave attack would be largely suppressed, because rivers would not carry the pebbly quartz to the coast. Not much quartz would reach positions where turbidity currents could carry them to the deep-sea floor, and the few that were carried by such flows would travel less far. In fact dry land would be smaller, higher, and more siliceous.

On the other hand, if quartz left the parent rocks in much smaller particles than actually happens, these would be carried more easily by rivers and marine currents and also suspended in the air. A higher percentage would be lost from the continents to the deep-sea. The continents would be less siliceous and thinner . . . and in general the protective cover of sand shielding more vulnerable material from weathering would be absent. Marine planation of the continents would be more active, because protective sands would have been wasted into the deep-sea. It thus appears probable that if quartz grains in granites were significantly bigger than they actually are, the continents would be smaller and steeper than we now find them. If granite quartzes were much smaller, the land would also be smaller but lower than at present.

On such details does the character and habitability of our planet depend.

Blowing in the Wind

Desert Landscapes

The desert is the Garden of Allah, from which he removed
all superfluous human and animal life, so that there
might be one place where he can walk in peace.

North African saying

A dense, stinging fog of low-flying sand grains wholly
obscured not only our cars but ourselves up to our shoulders,
while our heads stuck out against a clear blue sky. One after the other,
our feet dropped an inch as sand was scoured from beneath.
The whole landscape was on the move.

Ralph Bagnold, *Sand, Wind, and War*

JINNS

When sand moves under a gathering desert wind, it seems to take on a life of its own, to become a different form of matter—like a gas, like liquid nitrogen spilling and spreading, following the ground surface. Spraying off the crest of a dune, shimmering in the light, veils of sand race and ripple, spread and vanish, their place continually taken by the next gossamer sheet, dancing, playing, celebrating. Are these jinns, the spirits of the desert? The sight is beautiful and hypnotizing in the evening sun, but if the wind gathers speed, beauty rapidly vanishes as the violence and menace of a sandstorm grows. Suddenly, it seems as if the entire mass of desert sand has sprung from the ground to hurtle with the wind. On the surface, everything is moving, even the largest grains, rolling, tumbling, kicking smaller grains into the rushing current. The sky disappears, and the howl of the wind seems amplified by its cargo of sand. The air is filled with flying sand, unbreathable.

WIND

The desert is a stage on which wind and sand are actors and dancers, with everything else the backdrop. The sound of the wind is the sound of the desert. The winds have names: in North Africa, the simoun, the "poison wind," is searingly hot and dry, blasting everything in its path—it is the carrier of jinns. The sirocco, the ghibli, the khamsin, and the harmattan are the regional winds of the Sahara; the names themselves sound threatening. Out of those voices of the desert have come great religions and storm gods—Set for the ancient Egyptians, Jehovah, Baal, and Hercules, the ancient storm god of the Atlas Mountains. In his short story "A Passion in the Desert," Honoré de Balzac concludes: "In the desert, you see, there is everything and nothing. . . . It is God without mankind." The desert is the home of deities and strange and terrifying beings. Herodotus, in his wide-ranging but often fanciful descriptions of the arid lands beyond civilization, wrote: "There are enormous snakes there . . . donkeys with horns, dog-headed creatures, headless creatures with eyes in their chests (at least, that is what the Libyans say), wild men and wild women." The *nisnas* are mythical half-people of the Arabian deserts, running on their single leg, seeing with their single eye. Dorothy and Toto are blown to the land of Oz, which is isolated by the four forbidding expanses of the Deadly Desert, the Shifting Sands, the Impassable Desert, and the Great Sandy Waste, uncrossable by even the Winged Monkeys. The desert is, traditionally, the opposite of heaven, the character of hell for Dante. But today this has changed; as Robert Twigger observes in *Lost Oasis: In Search of Paradise,* the real lost oasis is the desert itself, "an oasis of light and contemplative beauty that replenished our inner reserves."

The winds are not only the voices of the desert, but are also often its cause. The major deserts of the world lie along the low latitudes north and south of the equator; these are the zones of subtropical high pressure, where the trade winds, having lost their moisture in the tropics, descend as dry air masses, precluding cloud formation and desiccating the land below. The Sahara and the Kalahari owe their origins to these winds. Elsewhere, the winds rise over mountain ranges, where they drop their moisture, to descend, dried out, on the other side. Rain shadow deserts result—in the western United States, east of the Sierra Nevada and the

Cascades; in Patagonia, east of the Andes; and in Central Asia, helped by the seasonal reversal of the monsoon winds, north of the Himalayas. Other major deserts lie along coasts where cold ocean currents, themselves driven in part by the winds, come to the surface, sucking the moisture out of the air. The Namib in Africa and the Atacama Desert in South America are so formed, the Atacama being officially the driest place on Earth—some parts had no rain from 1570 to 1971 (and when the rains did come, they caused devastation). The Gobi and Taklimakan Deserts of Asia are so-called continental deserts—places so far from sources of moisture that hot summers and cold winters generate desert conditions. Distinctions between desert types are not necessarily clear-cut, however, and many deserts result from a conspiracy of circumstances. For example, the Australian deserts—the Great Sandy, the Little Sandy, the Gibson, and the Great Victoria, among others—are influenced by the trade winds, rain shadows, and remoteness from ocean moisture.

But all deserts share one thing—extreme dryness. In strict terms, that dryness, the definition of a desert, is not simply a matter of low precipitation, but also takes into account the theoretical *capacity* to return whatever water is received from rain back into the atmosphere. It is the ratio between actual precipitation and the amount of moisture that could potentially be returned through evaporation and the activity of vegetation that determines whether a region is semiarid, arid, or hyperarid. Deserts fall into the last two categories, where the ratio is less than 0.2 (one to five) for arid regions and less than 0.03 for hyperarid. Much of the Sahara has the capacity to return to the atmosphere two hundred times the amount of precipitation that actually falls on it (a ratio of 0.005). Arid regions typically receive less than 200 millimeters (8 in) of rain per year, hyperarid regions less than 25 millimeters (1 in), and together they cover around 20 percent of the planet's land surface.

It should be mentioned here that the polar regions—deserts by definition and in reality the Earth's largest arid areas—are generally excluded. The entire continent of Antarctica receives an average of 50 millimeters (2 in) of precipitation per year, and there are sand dunes there, although on a far smaller scale than those of the classic deserts of the world.

SAND

Sand dunes cover only about 20 percent of the deserts (the sandiest being the Australian desert, which is half-covered in sand), but it is in the interplay between sand, wind, and ground surface that the dynamics of the arid landscape lie (Plate 9). The rest of the desert, the backdrop and the stage, is made up of mountains and badlands, the vast gravel plains of the *regs,* the ephemeral river valleys of the *wadis,* and the bare rock and boulder-strewn plateaus of the *hamadas.* The Sahara is the largest desert of all—it could comfortably cover the United States or Australia. The expanses completely covered in sand and dunes are the *ergs,* of which the Rub' al-Khali, the Empty Quarter, of the Arabian Peninsula is the largest, an area the size of France, covered in sand. The ergs are the image of the desert—vast, timeless, apparently lifeless seas of sand, the waves rolling and breaking in extreme slow motion, their spray carried by the wind—landscapes that have their own extraordinary beauty. Watch the opening sequence of the film *The English Patient* (sometimes referred to as *Gone with the Sand*), as the plane flies over the deeply shadowed dunes in the evening sun—it is a landscape of sensual beauty, nature as art.

The ergs of today are tangibly mobile, ever changing, but there are larger areas of ergs past that are now fixed by vegetation. Most of today's active sandy deserts are surrounded by vast stretches of old stabilized dunes, formed as the trade-wind belts and arid regions expanded in the cold, dry climate of the last ice age and immobilized as the climate changed. However, continuing shifts in the climate may bring these fixed ergs, granular reserves awaiting activation, back to life. We know that small changes in climate coupled with shifts in wind direction can create and remobilize dunes. The largest area of sand dunes in the Western Hemisphere covers around a quarter of the state of Nebraska. There, the Sand Hills were active and mobile between A.D. 1000 and 1200, formed originally from the debris of the glacial erosion of the Rocky Mountains. The hills were stabilized eight hundred years ago but have had episodes of reincarnation since: a long drought toward the end of the eighteenth century resuscitated dunes on the Great Plains, whose activity caused problems for the westbound wagon trains decades later.

Deserts, like rivers and oceans, are engines, with wind and sand doing the work.

Sand is supplied by the desert, deposited in the desert, and, occasionally, exported from the desert: a sediment budget again, on a huge scale. Where does all the sand come from? Essentially, all of it is provided internally, by the wind reworking ancient river and lake beds and by the sandblasting of the exposed rock, mechanical weathering, and erosion. Most deserts are topographical depressions, largely surrounded by higher areas, and so most of the sand stays in the desert, following the complex flow lines of the winds. A map of the sand highways of the Sahara reveals a number of points away from which the sand flows, sometimes in straight lines, sometimes in great whorls and arcs, weaving around the mountain massifs, pouring down through the ergs. A significant number of the flow lines end at the Atlantic or Mediterranean coasts, the export terminals for desert sediment.

Most of the sediment exported from the Sahara is not sand. As we saw in chapter 1, even the finest sand will settle from the air and not be carried huge distances in suspension; we will return to this later. The only Saharan sediments that can be carried great distances from the coasts are particles smaller than sand—silt and dust. And they are carried in enormous quantities. It is estimated that more than ten million dump truck loads of Saharan dust are exported each year, with the occasional giant "sand" storm carrying up to 100 million tons. The dust has been found coating the Greenland ice and is a probable source of nutrients to the rainforests of the Amazon. Florida receives perhaps 50 percent of African dust exports to the United States, and with the dust come microbes that have been accused of damaging coral and other marine organisms. The natural export terminals along the Mediterranean have resulted in sand being deposited in great quantities close to the coast, burying the ancient Roman ports and cities of Leptis Magna and Sabratha; the dust continues on to Europe, causing the "blood rain" reported throughout history.

These are huge global movements of sediment, but not of sand—although there is the occasional exception. Fuerteventura in the Canary Islands, today only 100 kilometers (60 mi) off the desert coast of West Africa, is famed for its sand dunes; the sand probably arrived at the height of the last ice age, when the lower sea level meant that the distance from the desert was much shorter and the erg much larger. Sand is today again being exported from Western Sahara—for artificial beach nourishment of the islands.

Most sand stays in the desert—where it occupies itself energetically. Aeolian activity—after Aeolus, the Greek god of the winds, whose domain was the desert—is the dominant process, and the only sound. Wilfred Thesiger, one of the great desert explorers and writers, described in *Arabian Sands* "a silence in which only the winds played, and a cleanness that was infinitely remote from the world of men." As the sun heats the desert in the morning, the air rises; more air rushes in to take its place, creating a rising wind, which gathers force during the afternoon. By the following morning, the sand of the erg is swept clean, the landscape rejuvenated by shifting sand. Every day, countless tons of sand are moved on by the wind. But how does this work? And why is it important to know?

While the word *desert* often connotes a landscape devoid of people, this is by no means true. Except in the most extremely hyperarid regions, people make a way of life in the desert, developing agriculture and communities around the desert margins. Upward of a billion people live in arid or semiarid environments, and their way of life is dependent on coexisting with sand. Regardless of whether or not, on the global scale, "desertification" is happening, the day-to-day struggle to keep houses, villages, roads, and fields free of sand is a real one. And just as modern coastal communities have developed in ways that were long considered unwise, so have we spread into arid lands that are not naturally our home and built commercial infrastructure there. If we want to understand climate change and changing aridity, we need to look into the Earth's past to interpret those changes over long periods of time. The only way to do that is to learn to read the record of ancient desert sediments through understanding those of today. It's important to know how the desert engine works.

The mechanical fundamentals of aeolian sediment transport and deposition are the same in many ways as those for the actions of water in rivers and oceans—air and water are, after all, both fluids. But they are very different fluids, and there are therefore some important and basic differences between the ways in which they work. Their densities, buoyancies, and viscosities are vastly different: the effective weight of a quartz sand grain in air, for example, is around two thousand times greater than it is in water, which means that its settling velocity is far more rapid. To start a sand grain moving, the wind speed must be considerably greater than that of a current of water. But, very influentially, wind, unlike water, can flow uphill.

THE SAND MAN

The journey of a sand grain tumbling in the wind is a complex one, and while many of the aspects of that journey are understood, there is much, again, that is not. The foundation of what we do know, and of the research that continues today, is entirely the result of the pioneering work of one man (of whom we have already heard)—Ralph Bagnold. Today's academic textbooks on sand transport often include advice along the lines of "for inspiration, read Bagnold (1941)."

Bagnold's early encounters with sand occurred after he was posted to Egypt in 1926. Shortly after his arrival in Cairo, he watched the first successful excavation of the Great Sphinx: "I watched the lion body of the Great Sphinx being slowly exposed from the sand that had buried it. For ages only the giant head had projected above the sand. As of old, gangs of workmen in continuous streams carried sand away in wicker baskets on their heads, supervised by the traditional taskmaster with the traditional whip, while the appointed song leader maintained the rhythm of movement" *(Sand, Wind, and War).* It was never an ideal place to construct one of the world's great monuments.

Arguments about the age and meaning of the Sphinx still rage—there are limitations to the wisdom of that which, according to the Sphinx's riddle, goes on four legs in the morning, on two legs at noon, and on three legs in the evening (the answer being humankind). However, its link with the building of the pyramids is clear, and King Khafre (or Chephren) was the likely builder. The Great Sphinx has spent most of its existence largely covered by the continuously drifting sand, with only its head, blasted and worn by that sand, protruding. The ravages suffered by the head confirm that the sand that buried the body has been its salvation, preserving it from abrasion. In spite of its role over the centuries as an inspiration for archaeologists, poets, travelers, and those who believe it was built by refugees from Atlantis, its life has largely been like that of an iceberg, demurely hiding its bulk beneath the surface.

It was not until early in the nineteenth century that serious attempts at excavating the Great Sphinx were made, but these were defeated by the enormous volumes of sand involved. Further efforts in 1858 and 1885 revealed a good part of the body and some of the surrounding structures, but these attempts were again

FIGURE 27. Ralph Bagnold as the young desert explorer in 1929 and as the distinguished scientist later in life. (Photos courtesy of Stephen Bagnold)

abandoned. The Great Sphinx had to wait until 1925 and the arrival of the French archaeologist Emile Baraize for its full glory to be revealed. Removal of the vast quantities of sand required eleven years of labor.

In watching the results of natural sand movements on a staggering scale, Bagnold perhaps had an inkling of the way in which his future would be intimately driven, grain by grain, by sand. The insatiable curiosity that had possessed him since childhood—from an early age he was "aware of an urge to see and do things new and unique, to explore the unknown or to explain the inexplicable in natural science" *(Sand, Wind, and War)*—would carry him through an extraordinary diversity of accomplishments until his death in 1990 at the age of ninety-four, still in full stride (Figure 27).

Bagnold was one of those larger-than-life characters, but he was also, unusually, deeply modest. In spite of his scientific achievements and accolades, he always regarded himself as an amateur. His obsession was with seeking the truth through well-designed scientific experiment and observation, unsullied by conventional wisdom or tradition. Peacetime soldiering, as Bagnold observed soon after arriving in Egypt, left time for extracurricular activities, and so he set off on adventurous excursions into the desert that, over the next twelve years, through 1938, would become more and more ambitious and extensive, earning him his place among the pioneering explorers of the desert. The Western Desert of Egypt in the 1920s was a little-known place outside the great oases; the maps were conspicuously empty, the words "limit of sand dunes unknown" providing the only description to be found in otherwise large areas of blankness. Bagnold regarded these words as a challenge, and whereas journeys by previous explorers had largely taken place on the backs of camels, he set out by car—first the Model T Ford and later the Model A, painstakingly adapted for the desert and the need to be entirely self-sufficient, particularly in water supplies. In determining that the giant dunes of the erg of the Great Sand Sea could be crossed by car, Bagnold made close and meticulous observations of the nature and quality of sand—early work on the physics of granular materials.

Bagnold's method of crossing dunes required careful selection of the right location and then full-frontal assault, flying up the side of the dune in a cloud of sand. He recognized that while the sand might be quite soft, the way in which the grains were packed together created a firm enough foundation for a truck to leave only shallow tracks from its deflated tires. Nevertheless, he recognized equally that the sand was "unreliable" and, depending on its location within the dune system, could contain "pools" of unconsolidated sand into which a vehicle could sink, instantly and deeply.

In battling his way across the Great Sand Sea, Bagnold realized that there were three fundamental questions about sand dunes and sand transport that had perhaps never been asked and certainly had never been answered:

1. What determines the distinct shape of the different kinds of dunes, and how do they retain that shape while moving inexorably across the desert?

PLATES

PLATE 1. A small selection of sand, demonstrating its astonishing variety: (from top) quartz sands from Florida, Sumatra, and Algeria, ooliths from Mexico, olivine from Tahiti, forams from Bali, and volcanic glass sand from the Galapagos. (Sands courtesy of Peter Newman; photo by author)

PLATE 2. Sands of South Africa, arranged and photographed by Loes Modderman. (Photo courtesy of Loes Modderman)

PLATE 3. The ingredients of beach sands: Isle of Wight (left), Provence (right). (Photos by author)

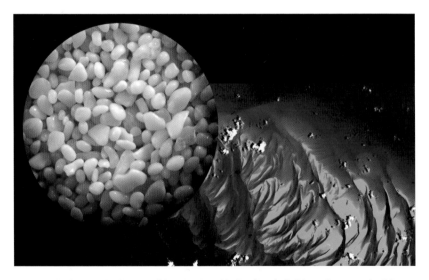

PLATE 4. A microscopic view of oolith grains from Mexico (inset). Ooliths make up much of the sand in the shoals of the shallow tropical waters of the Bahama Banks, seen in the satellite photo at right. The puffy "floating" objects in this image are clouds. (Inset photo by author; Bahama Banks image by NASA/GSFC/METI/ERSDAC/JAROS and U.S./Japan ASTER Science Team)

PLATE 5. *(facing page)* During the Loma Prieta earthquake of October 17, 1989, in California, sand volcanoes, or "boils," erupted in the median of Interstate 80 west of the San Francisco–Oakland Bay Bridge toll plaza. Ground shaking transformed a loose water-saturated deposit of subsurface sand into a sand-water slurry that vented along a fissure 7 meters (23 ft) long. (Photo by J. C. Tinsley, courtesy of U.S. Geological Survey)

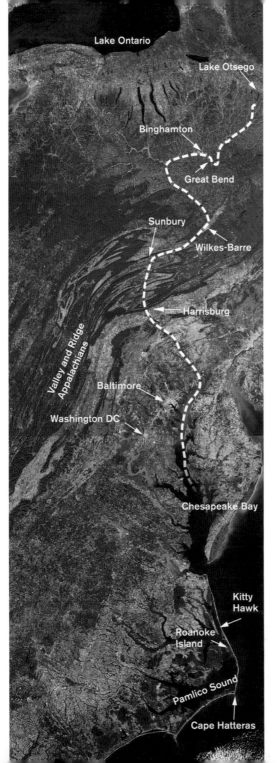

PLATE 6. A satellite view of the northeastern United States, showing the sand grain's journey down the Susquehanna River to the Chesapeake Bay—and perhaps beyond, to the Outer Banks. Note the plumes of sediment in the estuaries and around the Outer Banks. Locations referred to in chapters 4 and 5 are labeled. (Image by NASA)

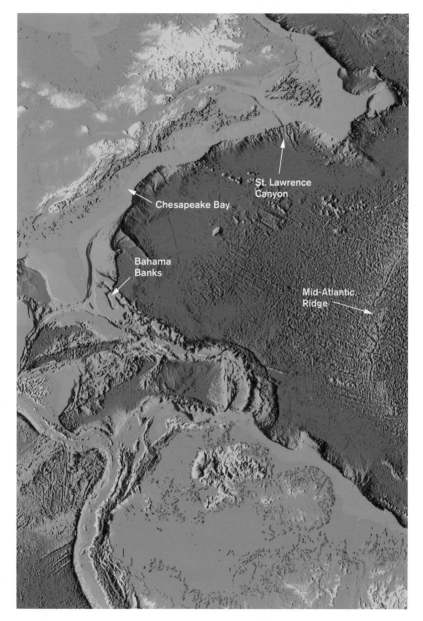

PLATE 7. Continental and oceanic topography: the deepest regions are dark blue, the highest are red. During periods of low sea level, the light blue areas beyond today's shorelines would have been dry land. (Image by NOAA, National Geophysical Data Center)

PLATE 8. Two sand dwellers: a ghost crab playing with sand, and a trumpet worm inside its shell. (Ghost crab courtesy of Robert J. Amoruso, www.wildscapeimages.com; trumpet worm courtesy of Kåre Telnes, www.seawater.no)

PLATE 9. Mountains, dunes, salt flats, and (sometimes struggling) vegetation of the Namib Desert. (Photo by author)

PLATE 10. A "street" of clear ground between dunes in the Western Desert of Egypt. (Photo by author)

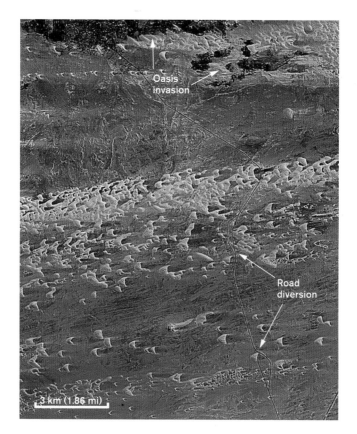

PLATE 11. Viewed from space, colonies of horn-shaped barchan dunes migrate southward (toward the *right* of the image) in the Egyptian desert; note the "babies" and the barchanoid ridges of coalescing dunes. In the bottom photo, one of the roads highlighted in the satellite image is being covered by an advancing barchan, the avalanching lee face moving toward the bottom left. (Top image by TerraMetrics; bottom photo by author)

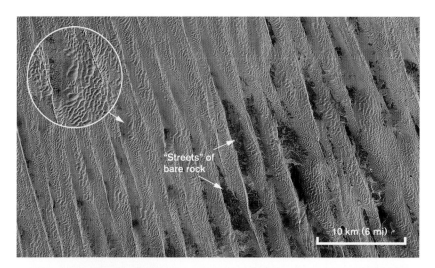

"Streets" of
bare rock

10 km (6 mi)

PLATE 12. Seif dunes of Egypt's Great Sand Sea, generally moving southward toward the bottom of the satellite image. The most active ridges are the pale lines on top of the broad foundations of sand, seen as sharp crests in the view from the ground (bottom). (Top image by TerraMetrics; bottom photo by author)

PLATE 13. Towering thicknesses of the Devonian Old Red Sandstone in Sofia Sound, East Greenland. (Photo courtesy of Peter Friend)

PLATE 14. Making a Vajrabhairava sand mandala. (Photo courtesy of the Huntington Archive, the Ohio State University, John C. and Susan L. Huntington)

PLATE 15. Examples of the sand art of Andrew Clemens. (Photo courtesy of the State Historical Society of Iowa, Des Moines)

PLATE 16. Burns Cliff, Endurance Crater, Mars, November 2004, recorded by the hardworking *Opportunity*. (Image by NASA/JPL/Cornell)

PLATE 17. Desert treasures. (Photo by author)

2. Why does sand gather itself into dunes at all, rather than being spread evenly over the desert floor? Why is sand self-accumulating?
3. How do individual sand grains interact with each other and with the wind to feed the dunes?

Being of an analytical and inquisitive nature, Bagnold decided that the only approach was to go back to basics, ignore the few theories that had been put forward to that point, and conduct some carefully designed laboratory experiments. He was provided space at Imperial College in London, where he proceeded to design and build a wind tunnel within which he could blow (or suck) sand grains under controlled conditions and carefully collect the quantitative data that he needed. It was an exquisite piece of apparatus, and the results were revolutionary. The experimental data, combined with the thousands of painstaking analyses of sand-grain sizes that Bagnold had assembled through endless sieving, culminated in *The Physics of Blown Sand and Desert Dunes,* finally published in 1941 while Bagnold was causing logistical havoc behind enemy lines in North Africa through the activities of his Long Range Desert Group. The book remains a masterpiece of scientific inquiry and analysis—it was used as a reference by NASA for planning Moon and Mars missions.

HOW SAND MOVES

Among Bagnold's key findings were that the basic laws of physics could be applied to sand movement, that the interaction between grains and the wind could be described quantitatively, and that the predictions that result could be tested by experiment and then verified by measurements in the desert.

When the wind blows over the desert floor, its flow is influenced by the nature of that surface, its roughness on all scales. Surface irregularities disrupt the smooth flow of air, causing turbulence and eddy currents. These in turn interact with the sand grains on the surface, which may be moved along or temporarily kicked up by the wind, which modifies its movement—a constant series of feedbacks between the wind and the grains. The act of moving sand grains removes energy from the wind and transfers it to the grains, which, colliding with their colleagues, transfer that energy in turn to them. The result is that, close to the ground surface, where most of the action is going on, the wind speed is reduced. There is a velocity gradient, whereby the wind speed increases with the height. Velocity gradients cause pressure gradients, and pressure gradients mean planes—and grains—can fly. What happens on a very small scale very close to the surface of the ground in the desert is critical to the grand-scale results.

A wind moving with the speed of a violent hurricane—perhaps 300 kilometers (190 mi) per hour—can pick up and transport pebbles, but typical winds deal with sand and smaller grains. Clearly, the wind speed needed to start sand grains moving depends on the size of the grains, but the minimum wind speed necessary to move the fine sands of the desert is around 16 kilometers (10 mi) per hour. Look closely at the sand dunes the next time you are at the beach: it's remarkable how even a light wind can nudge sand grains along the surface. This nudging is referred to as *surface creep,* but if the wind picks up a bit, it will lift grains very briefly off the surface. They fall back quickly, but when they do, they bang into other grains and kick them into the air. Very rapidly, the whole ballistic process gathers momentum, and in a moderate wind there will be a cloud close to the surface, comprising sand grains traveling by leaping and jumping, kicking off other grains as they land (Figure 28, top).

This is the process that we saw taking place in rivers in a less dramatic way, the impacts between grains being cushioned by water. In air, there is virtually nothing to dampen the impacts, and *saltation*—movement by jumping—is a violent business. Bagnold observed that a single high-speed saltating grain can move a surface grain more than six times its own diameter and two hundred times its own weight—saltation of fine sand can maintain movement of grains too large to be moved by the wind alone. This is fundamental. In a moving cloud of sand grains, the majority, perhaps 75 percent, are moving by saltation. As illustrated in Bagnold's diagram in Figure 28, the more pebbles there are on the surface, the more violent the grain impacts and the higher they bounce; if a saltating sand grain lands on a bed of sand, it splashes rather than bounces and much of the energy is lost. Recent work by Jasper Kok and Nilton Renno at the University of Michigan has added a further, influential, player to the game of saltation: electricity. As soon as the grains start moving, their collisions generate static electrical charges, which in turn facilitate the initial picking up of grains from the surface, adding significantly to the amount of saltating sand on the move but keeping it close to the ground.

Grain Paths over Pebble Surface

Grain Paths over Loose Sand Surface

0   5   10 Cm.

FIGURE 28. Bagnold's high-speed photographs of saltating sand grains in a wind tunnel (top); his sketches showing sand grains saltating more energetically from impacts with pebbles than with loose sand (bottom). In the photographs, the "wind" is blowing from left to right; a falling grain ejects another almost vertically into the flow. Ripples are beginning to form in the bottom image. (Images courtesy of Stephen Bagnold)

Bagnold's second question, as to why sand is self-accumulating, was answered by his observations of saltating sand. A blowing grain will continue to bounce energetically off pebbles but will splash into even a small pile of sand and have more difficulty moving on: the pile grows and a dune is born. At the same time, a rock-strewn surface between dunes will continue to bounce sand grains along. This accounts for the long "streets" of clear ground commonly occurring between dunes (Plate 10); covered with rock and pebbles but little sand, they allow a car access into the erg, but often alarmingly close off with no warning.

As with all forms of sediment transport and deposition, the laws of supply and demand prevail in the desert. As long as there is a supply of sand and sufficient wind, then the demands of the dunes will be satisfied. A wind will carry more sand over rocky ground, feeding the dunes, than over the dunes themselves—and the higher the wind velocity, the more sand it will carry. Because of the physics of what's going on, the ability of the wind to shift sand increases by the *cube* of the amount that its velocity exceeds the threshold velocity needed to start movement—exactly the same relationship seen in the moving fluid of a river. For the same grain size, a wind blowing at 16 meters per second (35 mph) will move as much sand in a day as a wind at half that velocity will in three weeks.

So, as we saw in chapter 1, size is critical. Larger grains that normally would not move are pushed along by the ballistics of saltation, and saltating grains bounce more energetically off larger grains than smaller—because of this interaction, there will be more mass movement if the sand is composed of different grain sizes (if it is poorly sorted) than if it's of a uniform size. But the range of size is limited by the processes of the wind, and the predominant size in that range is roughly 0.08 to 0.15 millimeters ($\frac{1}{300}$ to $\frac{1}{170}$ in): much of the sand in all of the deserts of the world is in this size range.

Even in a strong wind, gravity wins and the height of the cloud of saltating grains doesn't reach your knees (although it does a painful job of sandblasting everything lower). But turbulence and eddy currents can pick up fine grains and carry them in suspension for some distance, and so the total height of the cloud of whirling sand may well be up to your shoulders, while your head is in clear air.

In a particularly violent sandstorm (and I can vouch for this), the air is charged with flying sand grains above head height and surprising quantities of sand can be

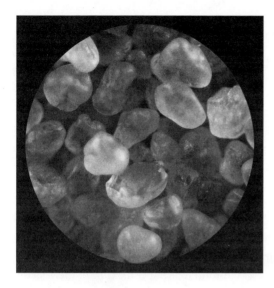

FIGURE 29. Rounded sand grains of the Sahara. (Photo by author)

driven into uncomfortable places. Typically, suspended sand grains remain aloft for only a matter of seconds, while dust can be blown around for years without ever coming back to earth. However, again as we saw counterintuitively in chapter 1, smaller grains are more difficult to *start* moving than larger grains because of their being protected by larger grains, the way they can pack themselves together, and, with even the smallest amount of moisture, the effects of surface tension between the particles. It takes a fair wind to start dust moving, but once it's in the air, it's gone, leaving the desert to the sand.

The sheer violence of the endless impacts between moving sand grains has another important consequence: it knocks off any rough edges on the grains, even the hardest ones, and they become more *rounded*. Our sand grain journeying down the Susquehanna traveled a long distance, rolling, tumbling, and bouncing along the riverbed, but because of the cushioning effect of the water, it suffered little damage. In the wind, however, there is no cushioning, and a grain that started off with rough edges will be abraded and become rounded very effectively. Once a grain has become relatively round and smooth, it is resistant to further abrasion—the process of making the grain *smaller* is a very slow one for quartz sand. Blowing in the wind is by far the most effective way of making angular sand grains rounded; desert sand is typically rounded and smooth (Figure 29), and this is diagnostic, as we shall see, for the forensics of ancient desert sandstones.

Bagnold's meticulous work answered his second and third questions, but what about the first, the life and behaviors of the dunes themselves? As he observed in the introduction to *The Physics of Blown Sand and Desert Dunes,* in the dunes, "instead of finding chaos and disorder, the observer never fails to be amazed at a simplicity of form, an exactitude of repetition and a geometric order unknown in nature on a scale larger than that of a crystalline structure." Why?

MOVING MOUNTAINS

Lie down on the desert sand, across the direction of the wind, and sand will gradually drift up against you. In a modest wind, your body weight in sand may well accumulate in an hour; in the gale of a sandstorm (remember the cube of the velocity rule), you may be buried in a ton of sand. Burial by sand makes for a good story, and there are many. The history of Arabia and North Africa features tales of armies dispatched to fight the sand and the wind, only to be overwhelmed. One such story that contains real characters and therefore might be closer to truth than myth is told by Herodotus in his history of the Persian wars, written around 450 B.C. Cambyses, the Persian conqueror of Egypt, was being given a great deal of trouble by the Ammonites, the guardians of the Oracle at Siwa (whom Alexander would later consult). In 525 B.C., Cambyses assembled an army and began the march across the desert to Siwa. The army departed the oasis of Kharga and was never seen again: forty thousand men, together with their animals and equipment, "lost at sea," buried in the sand.

In Mark Twain's *Tom Sawyer Abroad,* Tom, Huck Finn, and Jim take off on a Jules Verne–like voyage in "the boat," a science-fiction hot air balloon. Sailing over the Sahara, they watch a camel caravan plodding over the dunes below them:

> Pretty soon we see something coming that stood up like an amazing wide wall, and reached from the Desert up into the sky and hid the sun, and it was coming like the nation, too. Then a little faint breeze struck us, and then it come harder, and grains of sand begun to sift against our faces and sting like fire, and Tom sung out:

"It's a sand-storm—turn your backs to it!"

We done it; and in another minute it was blowing a gale, and the sand beat against us by the shovelful, and the air was so thick with it we couldn't see a thing. In five minutes the boat was level full, and we was setting on the lockers buried up to the chin in sand, and only our heads out and could hardly breathe.

Then the storm thinned, and we see that monstrous wall go a-sailing off across the desert, awful to look at, I tell you. We dug ourselves out and looked down, and where the caravan was before there wasn't anything but just the sand ocean now, and all still and quiet. All them people and camels was smothered and dead and buried—buried under ten foot of sand, we reckoned, and Tom allowed it might be years before the wind uncovered them, and all that time their friends wouldn't ever know what become of that caravan.

Realizing that "the boat" (apparently defying the laws of physics) contains several tons of "genuwyne sand from the genuwyne desert of the Sahara," they consider the commercial potential of putting it into vials and selling it back home (presumably to early arenophiles), "because it's over four million square miles of sand at ten cents a vial." But the scale of the operation—and that of the duties they would have to pay—leads them to abandon the idea.

Because of the sheer scale and inaccessibility of the sand seas, the way in which they and their dunes formed was completely unknown before Bagnold got to work. An ability to see the relationships between minute and gigantic scales was part of Bagnold's genius: he simply took his work on the behavior of individual and small groups of sand grains and applied it to the larger scale. Until then, sand dunes had simply been thought of as very large ripples, for ripples form on the surface of desert sand just as they do at the beach or on a channel bar.

We have seen that ripples formed by water are enigmatic features, not fully understood, and the same applies to those made by wind. Bagnold suggested that small initial irregularities of the ground surface interact with the jumps of saltating grains to form ripples, with the distance between the crests of the ripples being directly related to the lengths of the jumps. Grains impacting the windward side nudge other grains of a certain size over the top, and these are then sheltered on the lee side while coarser grains become marooned on the crest of each ripple. But profound complexity can characterize even small features: while this process fits many

FIGURE 30. Sand ripples, small and mega. (Photo by author)

of the facts, it is not consistent with all of them, and other mechanisms have been suggested. It is certainly true that under the right circumstances, very large ripples, termed, appropriately, *mega-ripples* or *sand waves,* can form. The wavelengths of these can be 25 meters (82 ft), and they can be 30 centimeters (1 ft) high; they make for appalling driving conditions, very much like moguls on a ski run. Figure 30 illustrates small-scale ripples forming on the flanks of mega-ripples. The latter seem to form differently than the smaller ripples; the grain size is larger and the internal structure is different, and they are common where the wind is funneled violently between rock outcrops. But however ripples, large or small, form, they are not simply small dunes. The largest mega-ripples are smaller than the smallest properly formed dunes, and no gradual transitions are seen. So what constitutes a properly formed dune?

It doesn't matter whether you are in the Sahara, the Namib, the Sonoran, or any of the world's deserts, there are distinct types, or "species," of sand dunes that can form, their shapes depending on sand supply and wind directions. There are species that form transverse to the wind direction and those that form parallel to it, those that form in response to complex winds and those that result from the interaction with topography and vegetation. And, of course, there are mixtures of all of the above—the classification of dunes is an inexact science. In fact, even today, the understanding of how dunes form and behave has real limitations. No one has ever *made* a dune, even a small-scale laboratory dune. Piles of sand, yes, but they never evolve into a working, moving, living sand dune. Computer modeling helps, but much of what we understand today is still based on Bagnold's fascination with dunes. Plates 11 and 12 illustrate the most common "species": *barchan* and *seif* dunes.

Perhaps the simplest and the most elegant dunes are the individual crescent- or horn-shaped barchans (from the Arabic word for a ram's horn). Many of the dunes in the Nebraska Sand Hills are barchans, as are those in the Great Sand Dunes National Park in Colorado. Bagnold saw in barchan dunes some of the attributes we ascribe to organisms, living things with inherent shapes and behaviors:

> Here, where there existed no animals, vegetation, or rain to interfere with sand movements, the dunes seemed to behave like living things. How was it that they kept their precise shape while marching interminably downwind? How was it that they insisted on repairing any damage done to their individual shapes? How, in other regions of the same desert, were they able to breed "babies" just like themselves that proceeded to run on ahead of their parents? Why did they absorb nourishment and continue to grow instead of allowing the sand to spread out evenly over the desert as finer dust grains do? *(Sand, Wind, and War)*

Barchans are the simplest form of transverse dunes, forming at right angles to the wind direction and depending on that direction being constant, without seasonal variations. It is this constancy of the wind that leads to the beautiful symmetry of the barchan (whose form is also seen in underwater dunes where currents are constant). An initial patch of sand grows, as we have seen, by attracting saltating grains. As it grows, developing into a mound of sand, it begins to react with the wind that is forming it. Grains build up a gradual slope on the windward side, while a wind shadow develops on the lee side. A crest forms above the steepening lee slope, and the dune develops its typical asymmetrical profile. As sand is carried up off the windward slope, avalanches (the natural behavior of granular materials) start pouring down the steeper lee side, the *slip face,* and so the dune migrates. In a strong wind, the avalanching is continuous and dramatic, and the crests of the dunes are shape-shifting folds, curtains of moving sand, compared by Bagnold to the gale-driven motion of the mane of some huge animal.

Figure 6 showed sand avalanching in a liquid-like action down the slip face of a dune, like a still from a movie of masses of moving sand pouring down a slope, peeling off across the face. This process of avalanching not only results in the dune migrating, but produces huge-scale cross-bedding in its internal structure, a diagnostic feature for reading the records of deserts from the Earth's past.

Around the edges of the dune, where there is more bouncing and less splashing, the rate of sand transport is greater, and so the flanks of the dune extend into the characteristic "horns" of the barchan shape; the dune moves in the direction in which the "horns" are pointing. These dunes can grow up to 300 meters (1,000 ft) wide and 100 meters (330 ft) high, but smaller ones are common—isolated individuals marching in a herd across the desert floor. And march they do—at rates of up to 30 meters (100 ft) every year. During a far-reaching 1930 expedition, Bagnold and his colleagues were making their way across the Selima sand sheet of Northern Sudan, a flat expanse of sand the area of Wyoming. The only shelter they could find for the night was a single barchan dune, apparently the scout out in front of the rest of the herd. They camped on its sheltering lee side, out of the wind, and in the morning moved on, leaving their empty cans to be buried beneath its sand. This was Camp 18, whose location was carefully determined astronomically; the vital statistics of the dune were also carefully measured.

In February 1980, an American geologist, Vance Haynes, was traversing the Selima sand sheet and, in the middle of nowhere, happened upon a small pile of empty tins—Bagnold's rubbish. He confirmed their identity with the brigadier and noted that, though Bagnold's idea of the tins being covered by the sand might have worked for a while, the sand had continued to move on and had now left the pile in its wake—the stern of Bagnold's barchan was now over 150 meters (490 ft) away. Haynes made careful measurements and set out markers that he monitored for the next seven years. In the end, thanks to Bagnold's rubbish, he had measurements of the dune's movement over a period of fifty-seven years. It had moved 7.5 meters

(25 ft) per year. Individual barchans move at different rates, the smaller ones often being more skittish than their larger relatives. A small one will often rear-end a larger one and grow; the larger one in front is starved of sand, diminishes in size, but picks up speed, eventually detaching itself and moving off. This leads to the impression that the original small dune has merged with and passed entirely through the larger one, but modern computer modeling of the process shows that this is not the case. Occasionally, the horns of a barchan will separate from the body of the dune, grow into small self-contained dunes, and move off—the reproduction mechanism that Bagnold referred to. Very commonly, because of their different rates of advance, there are large pile-ups of barchans, forming a complex series of *barchanoid* ridges, transverse to the wind but complex in their flowing shapes and continuously shape-shifting crests and slip faces.

Where there is more than one significant wind direction, different species of dunes form, typically long linear forms, in reality often sinuous, joining and bifurcating, sometimes building up into huge relatively stable platforms (called "whalebacks" by Bagnold), on the top of which elongate *seif* ("sword") dunes are the active players. These dunes can be gigantic, both in height and length; all require a constant and plentiful supply of sand. Linear dunes are by far the most common type in the Australian deserts, where many are now stabilized, but they are common to all the great ergs of the planet.

Linear and seif dunes form broadly parallel to the dominant wind direction but seem to result from seasonal, often stormy winds from a different direction; these set up large-scale eddies and vortices between the dune chains, clearing sand out of the intervening "streets" or redistributing grains from the broad foundations of sand into the elongated dunes (Plate 12). Transport of sand alternates direction, building up downwind along one side of the dune for part of the time and then on the other, extending the dune in the average direction of the winds: the dune gradually migrates. This is also true of *star* dunes, which show central peaks with radiating arms and multiple slip faces; they often form into networks on, again, a broad plinth of sand and may result from multiple wind directions.

Once topography—cliffs, wadis, mountains—becomes involved, dune shapes and processes can become yet more complex and mysterious. The Great Sand Dunes National Park is a spectacular example. The southwesterly winds pick up sand from the old flood plain of the Rio Grande and transport it until they run into the towering obstacle of the Sangre de Cristo Mountains. Funneling into an embayment in the range, the winds lose energy and have to dump their load in order to gain altitude and cross the mountains—hence the dunes. However, storm winds come from the northeast, and the combination of variable, swirling winds and mountain topography creates a mass of complicated, merging, ever-changing dune types, more varied than those we have mentioned here. The result is a landscape of staggering complexity and beauty.

But for communities living with the threat of ever-moving mountains of sand, that beauty has limitations.

SURGING SANDS

The beleaguered character in Kobo Abe's *Woman in the Dunes* may be fictional, but the sand women of Arawan are a fact. The once-thriving oasis north of Timbuktu used to serve the camel caravans from the salt mines farther north, but for a long time now the visitors have been dunes rather than traders. And the women of Arawan shovel sand, every day when the wind is not blowing too hard. It's not that they haven't always lived with sand, being, after all, in the middle of the Sahara; but in the 1960s and early 1970s a series of devastating droughts destroyed much of the vegetation, and the sand began to move. The mosque was buried, trees were engulfed, and the wells had to be reexcavated. Dunes encroach on the roofs of houses, the owners keeping only the doorway clear. The sand women are professionals, often paid in rice or sugar to shovel the surrounding dunes with the same energy and futility as Abe's "woman in the dunes."

Arawan is by no means alone in suffering this invasion. In towns and villages in Algeria and elsewhere in the Sahara, houses engulfed by sand are abandoned, to be reoccupied, perhaps by the next generation of the family, after the sand dune has moved over and onward. Fields and palm groves are treated the same way—but the palms will be dead and new ones will need to be planted. Even in the great oases of Egypt, isolated herds of barchans march relentlessly over oasis villages. Huge amounts of money have to be spent keeping roads and railways clear of sand or rebuilding them. Outside the oasis town of Kharga in Egypt's Western Desert, a family group of perhaps nine individual barchans plays havoc with the main road (Plate 11). The highway has had to be rebuilt along different routes five times in the past forty years; lines of telegraph poles, only their tops protruding from the sand, mark old routes. Today, a dune threatens the newest segment of road but has moved past an older one, which can be brought back into service.

Encroaching sand is a constant menace to agriculture, oases, and infrastructure in all arid regions of the world. Even late in his life, Bagnold found himself "in some demand as a sage on the subject" *(Sand, Wind, and War)* and was summoned to various parts of the Middle East to assess the threat posed by migrating dunes.

The threat, of course, is not new. Populated oases have been dealing with the problem of sand for thousands of years. But those problems seem to be getting worse, and waves of sand are breaking over villages, towns, fields, and infrastructure every day. So how to deal with them? Shutting off the supply at its source is as impracticable as it is for the oceans, so intervention and defense are the only options. The traditional defense has always been building fences of date palm fronds to capture the encroaching sand, a temporary solution at best. Surface coatings have also been used to stick the grains together and stop sand movement. Water can be sprayed on the surface of a dune, but water is a scarce and valuable commodity in the areas where such problems arise. The dunes at Kharga have been partially covered in tarmac—unsightly and expensive. Since there tends to be more oil than water in desert regions, oil has been sprayed on dunes—expensive, polluting, and potentially toxic. Today, there are specialty chemicals that avoid the pollution and toxicity, but they remain far from cheap.

Other defenses have been devised that use what seems to be an innate knowledge of the physics of moving sand. In the early twentieth century, engineers constructing Peru's railways spread pebbles and gravel over threatening dunes, a solution based on the natural process of *deflation:* wind erosion lowers the surface of the lands, winnowing out sand from between larger fragments until all that is left is a layer of pebbles and gravel, creating a desert pavement that resists any further erosion. The solution also drew on the process of *saltation.* As we know from the principle of bouncing versus splashing, sand blown across pebbles will saltate with more energy, and so it moved on across the railway tracks rather than stopping at the dune—and the dune was starved of sand.

Many *physical* remedies along the lines of palm frond fences have been tried by different communities, largely on an ad hoc basis, but Jean Meunier, a retired French agriculture teacher, was the first person to attempt a coordinated approach using the basics of blown sand and desert dunes. In the early 1990s, he applied his interventionist thinking around Nouakchott ("place of the winds"), the capital of Mauritania, a city whose growth has put it on a direct collision path with the surrounding dune fields. Meunier's approach was simple: harness the wind energy itself to destroy the dunes. To do this, he developed two types of fencing structures, semipermeable and impermeable. The semipermeable he would place on a dune upwind from a threatened structure. The wind could pass through the barrier, but the sand grains could not. On the downwind side, the wind, relieved of its load, would have the energy to scour and *remove* sand already deposited. It worked: for example, he saved a school from burial this way. His other method was to use solid, impermeable barriers to divert the wind and cause it to attack a dune directly. Most of the dunes threatening Nouakchott are linear, piled up by seasonally alternating winds from the northeast and northwest. Meunier would build what he called *guillotines,* V-shaped barriers, each arm at right angles to one of the winds. They forced the wind around and over them, disrupting the natural flow and causing large eddies that scoured the sand. The dune would be split in two in a matter of weeks, its natural form destroyed, its ability to move crippled. He developed a similar method for self-destruction of a barchan. As Meunier remarked: "If you destroy the shape of the dune, it becomes simply a pile of sand and can no longer migrate" (Zandonella, 2003).

Once he had reduced a dune to a mere pile of sand, Meunier then applied the coup de grâce—he vegetated it. Planting vegetation has long been an approach used to stabilize dunes, with varying degrees of success. The kinds of grasses used to plant coastal dunes are ineffective in the desert—their roots are too shallow for the lack of moisture, they become quickly buried, and their leaves are inadequate for trapping sand. Bushes and trees are more effective in the desert (as long as the planting pattern does not generate a wind-funneling effect, which only exacerbates the problem); however, relatively small areas need very large numbers of them. Meunier used small desert shrubs and an innovative watering system that encouraged the roots to rapidly reach depths where there was sufficient natural moisture.

Meunier's methods were initially greeted with skepticism, but his success brought collaboration, and after ill health forced him to leave the Nouakchott project in 2002, his methods, supported by new computer modeling of dune dynamics, continue to be the focus of activities by nongovernmental organizations.

If one were to try to guess which country is most desperately in need of innovative solutions to combat the surging seas of sand, one might think of one of the countries of the Sahara, or possibly the Middle East. But that country is China, not only because of the scale of the problem, but because of the scale of the threatened population and infrastructure. More than a quarter of the total area of China, over 2.5 million square kilometers (a million sq mi)—about four times the area of the state of Texas—is covered by deserts. Four of these—the southern Gobi, the Alashan, the Taklimakan, and the Tengger—are more or less connected to form the vast arid interior. The total area of China's deserts is growing at around 200 square kilometers (80 sq mi) every *month,* and every year tens of thousands of tons of sand and dust are blown into Beijing. China's capital has always suffered from dust storms, helped again by the ice ages, when grinding glaciers wore rocks down to flour, technically known as *loess,* which, once airborne, blankets huge areas for long periods of time. But Beijing's dust storms are turning into sandstorms. It's not necessary to travel to the Gobi Desert to find encroaching sand; it's a mere hour's drive out of Beijing. The Great Wall, built to defend against invaders from the west, is proving no match for the onslaught of sand: whole sections are being destroyed by the storms.

In the village of Longbaoshan, dunes are consuming the houses, and digging has become a way of life. In the region where forests and lakes once provided the hunting grounds for emperors, sand dunes move across the landscape at 20 meters (65 ft) per year. Entrepreneurs have benefited from tourists and filmmakers traveling to Longbaoshan from Beijing in quest of desert landscapes, but the villagers' narrative could, again, have been taken straight out of *Woman in the Dunes:* "Sometimes I dream of the sand falling around me faster than I can dig away. The sand chokes me. I worry that in real life, the sand will win" (Gluckman, 2000). In a few years, Beijing will be facing not only the airborne assault of sand, but also its ground troops.

In the heart of its deserts, moving sand has long been a challenge for China.

Whenever roads and railways are built, sand clearance becomes an ongoing necessity. Vast stretches of rail tracks and roads are constantly blocked by sand. The Shapotou Desert Experimental Research Station, on the edge of the Gobi, has an international reputation in desert research. The defense methods developed there vary. Grids of straw laid out across the dunes have been successful saltation inhibitors, at the same time allowing vegetation to be reestablished. Revegetation—on the huge scale characteristic of so many Chinese projects—is the focus of current efforts. Villagers, schoolchildren, farmers, and herdsmen all receive incentives to plant trees, shrubs, bushes, and grasses—as fast as possible. What has been called the "Green Great Wall," an arc of vegetation bordering the entire southern margin of the Gobi Desert, is already well on its way to completion. Gigantic plantations designed to protect Beijing are also underway. But all this consumes huge quantities of water that is no longer available for other uses. Is this the solution?

Today's inhabitants of villages like Longbaoshan can point to dusty, bare hillsides and valleys that they remember being verdant. It is estimated that firewood collection, excessive grazing, and overcultivation account for close to 90 percent of recent "desertification." Wind patterns can change on a human time scale—in the Sand Hills of Nebraska, for example—and, on a longer scale, climate can change, but for many areas of the world where sand is threateningly on the move, it is not nature but humans that are the cause.

THE GRANULAR ORCHESTRA

The Gobi is not the largest desert in the world, nor the sandiest, but nevertheless it is vast and difficult. The first Western traveler to penetrate the Gobi was Marco Polo; in his description of his journey, he mentions some of the many spirits of the desert, Asian jinns, among which are those that create sounds like musical instruments, a terrifying and mysterious drumming in the dunes. Ninth-century local records also talk of the "Hill of Sounding Sand," spontaneously emitting noises or doing so under the inducement of local villagers sliding en masse down its slopes as part of their ritual worship of the dune.

Such stories are common in arid lands—and in the great tracts of coastal dunes. Sand dunes around the world spontaneously give voice, emitting a wide range of sounds, booming ominously or simply singing to a weary traveler. Charles Darwin described the Chilean sand hill known as El Bramador, "the bellower," and noted the "chirping" sounds made by horses' hooves in the sand. Guy de Maupassant, traveling through Algeria, described how "somewhere, close to us, in an undefined direction, a drum was beating, the mysterious drum of the dunes." Wilfred Thesiger and many other desert explorers were startled by the chorus of the dunes, ascribed by their terrified guides to jinns, sirens, the bells of buried churches, or the drummer of death. And finding himself having to shout in order to converse with his colleagues during one such performance, Bagnold took a particular interest in "the weird chorus."

The repertoire of sand grains is extensive and varied. Dunes have been described as booming, roaring, thundering, whining, squeaking, and singing. Comparisons have been made to the sounds of a drum, a zither, a horn, a cello, a trumpet, a didgeridoo, bees, a foghorn, and low-flying aircraft. Dozens of dunes worldwide have voices, from the coasts of the Scottish islands to the Kelso Dunes of California, as does every deep desert. Sand Mountain, Nevada, emits a low C, while dunes in Chile have been noted to play an F, and dunes in Morocco a G sharp. The phenomenon—referred to as "dune tunes," "the sand of music," "music of the spheres"—has considerable scientific and popular appeal. But what causes it?

Scientific opinion varies widely, but it is clear that, as the ancient Chinese could induce sound by running or sliding down the face of a dune, it is caused by sand movement, or avalanching. Bagnold experimented with inducing sound both in the desert and in the laboratory with different types of sand under different conditions. My own experiences of attempting to seduce a dune into singing to me have been extremely frustrating. I have watched countless natural avalanches cascading in complete silence. Since the orchestra is often reported to perform most commonly at night, I have spent some time perched on top of a slip face in the moonlight, trying various devices (Bagnold succeeded with pencils, bottles, and his hands), with no success at all. Then finally, as I ran and plowed down the face of a Western Desert dune in broad daylight, suddenly through the avalanches at my feet emerged creaking sounds that built in resonance to a satisfying though modest booming sound.

The physics of granular materials—avalanching and the interactions of grains in motion—has led to significant research efforts into sand sounds. But there are several camps, each with a particular theory and often at quite dramatic personal loggerheads with one another. A fundamental division arises from whether the phenomenon results from behaviors of individual grains or the large-scale structure of the dune itself. If the sounds come from individual grains, what is the importance of sorting, rounding, dryness, polishing, and mineral coatings? Or does the avalanche set up a standing wave that resonates with the internal structure, effectively turning the dune into a loudspeaker? Is it the dilation and compression of the mass of grains, or does it have something to do with "triboelectrification," static electricity charges among the grains? What are the characteristics that result in high-frequency squeaking versus deep booming? Why does a particular dune change its tune from summer to winter?

Data from nature and the laboratory can be found to support essentially every theory. For the moment, dune voices remain among nature's most elusive and haunting secrets.

DENIZENS

The intrinsic acoustic properties of sand may be entertainingly mysterious to us, but they are vital to some of our fellow creatures. The desert environment, described as a "mineral world" by Bagnold and as an "iron landscape" by Antoine de Saint-Exupéry, will often seem to be devoid of life, but it is far from inorganic. Even in the depths of the hyperarid desert that supports no human life, when you wake up in the morning, the sand around you will be patterned with a network of tracks, large and small, left by inquisitive creatures of the night. Once, when photographing dune avalanches, I looked down at my hand and perched there was that stalwart of suburban gardens, a ladybug.

One such creature—and one that you would prefer not to encounter—is the sand scorpion. Like many desert residents, the scorpion solves the problems of extreme temperatures by burrowing and then listening to the sand. The acoustic background in the desert is blank, save when the wind blows. To us, there is utter silence, but against this apparent silence the blind scorpion has developed an exquisitely sensitive detection system. Drop a sand grain a short distance from it, and the scorpion will react. The movement of a small insect, the scorpion's dinner, will set up microscopic low-frequency acoustic waves in the sand that are instantly detected by sensors at the ends of the scorpion's eight legs. Not only does it detect the insect's presence, but it also immediately knows in which direction it can be found—the sound reaches the legs closest to the insect first, those farther away a millisecond later. And that millisecond is enough for the scorpion to whirl and pounce. These scorpions are also very good at detecting airborne vibrations and have a sense of smell orders of magnitude better than ours—they can sniff a sand grain and detect a mate.

There are two basic challenges for life in the sand: temperature and moisture. Many desert denizens, like the sand scorpion, solve the first by burrowing to levels where the temperature is stable, protected from the surface heat. Desert mammals, reptiles, and insects demonstrate some remarkable adaptations to help with digging in the sand. Australia is the home of twenty species of burrowing frogs, mice that hop—and the itjaritjari. The itjaritjari is a strange little marsupial mole that lives throughout Australia's deserts. It seems to have no eyes or ears, but put it down on the sand and it will disappear, as the Aboriginal people say, "like a man diving into water." No one knows how the itjaritjari navigates or senses, but it has a hard nose and front feet well adapted to excavating, the back ones being webbed to help push through the sand; the young are carried in a pouch that cleverly faces backward so as not to fill up with sand. Lizards too have different front and back feet designs for efficient excavation, and often very short legs. The sand skink, a lizard also known as the sand swimmer or the sand fish, seems to have taken a kind of nanotechnology approach, the microscopic design of its scales reducing friction and minimizing abrasion as it swims effortlessly beneath the surface of the sand for much of its existence.

Lizards have transparent eyelids that protect their eyes when burrowing or moving around on the surface. Some geckos have notched eyelids that interlock to keep out the sand—and a long tongue to clean them off. Snakes, however, have no eyelids; they can't blink or close their eyes. Sidewinders and some other snakes have enlarged horns above their eyes, so that when they burrow, the pressure of the sand against the horns effectively closes their eyes protectively. And the sidewinder, as its name signifies, is, together with other species, a snake that adopts an unusual form of locomotion across the surface of the sand, arching, twisting, and propelling itself in a series of sideways jumps. This is a highly effective way for a snake to move over granular material and, by keeping segments of the body off the hot ground, may also help it avoid overheating. Sand skinks, sidewinders, and sand vipers are also exquisite examples of sand camouflage. The viper will slither into the sand until only the top part of its head is exposed; essentially invisible, it then waits for its dinner to pass by.

Life in the sand desert is carried on largely underground. Ants carry individual grains from their nest to build a protective rampart around the entrance; wasps cover their immobilized prey with sand to hide it while they excavate a nest, lay an egg, and then bury the prey as food for the larva; and beetles inter themselves by stimulating small avalanches of sand. Fennec foxes, whose huge ears act as coolers and whose hairy feet help them travel through soft sand, can burrow 10 meters (more than 30 ft) into a dune. For the ancient Egyptians, the scarab beetle, emerging from beneath the desert sand, was a potent symbol of rebirth.

Underground living may solve the temperature problem, but it does little for the challenge of getting something to drink. There are beetles in the Namib that seem to use sand to solve the water problem. The same cold offshore current that dries out the coastal Namib also creates fogs that roll in over the dunes. Button beetles excavate furrows in the sand at right angles to the fog-carrying wind; the furrows disrupt the airflow, causing eddies from which moisture condenses and the beetles can drink. There is often dew in even the driest desert, but for many creatures it is simply consuming each other that provides sufficient moisture.

And then, of course, there are the ships of the desert, *Camelus dromedarius* and *Camelus bactrianus*. With a third sand-wiping eyelid, luxurious sand-filtering eyelashes, ears filled with sand-inhibiting hairs, closeable nostrils, and broad feet for sand walking, they are the ultimate desert sand machines. Their height also keeps their heads well above all but the worst sandstorms. Despite their ugliness, their foul smell, and their cantankerous nature, camels are highly valued and remain the means by which the Tuareg, the Bedouin, and all nomadic tribes of the Sahara and the Arabian deserts make a living. Today, there are around twelve million dromedaries in the world, all of them domesticated; their wild ancestors are long extinct. It was the camel that allowed Western eyes to be opened to the desert, that provided the means of access for the great early journeys of desert exploration. These journeys began with the epic travels of Marco Polo and Ibn Battuta in the thirteenth and fourteenth centuries and continued with the exploits of Wilfred Thesiger and T. E. Lawrence, the search for Timbuktu, and the intrepid wanderings of various Victorian British women.

THE INFINITE PRESENCE

Through these journeys, the desert was revealed and many of its myths dispelled, but it never lost its sense of vastness, its mystery, and its power—over human flesh and over the human imagination. It has created mystics, heretics, hermits, saints, prophets, and philosophers. It remains a place where god is, and where humans are only temporarily—"Look on my works, ye Mighty, and despair!" reads the inscription on the crumbled sculpture of Rameses the Great in Percy Bysshe Shelley's "Ozymandias." It is a place where human senses are distorted or redundant—for centuries, blind desert guides have navigated by the smell of the sand. It is a place to which we are drawn but within which we do not belong. In *The Sacred Desert: Religion, Literature, Art, and Culture,* David Jasper describes "a landscape that both kills and redeems and is absolutely indifferent and pure. It is never and always." Edmond Jabès, a Jewish writer and poet who was raised in Egypt but fled to Paris in 1956, was one of the great twentieth-century philosophers of the desert, merging the idea of the book and the process of writing with the mystical qualities of the desert: "We saw later how the book was but the letters of each word and how this alphabet, re-used thousands and thousands of times in different combinations, slipped through our fingers like grains of sand. Thus we became aware of the infinite presence of the *desert.*" And: "What is a book but a bit of fine sand taken from the desert one day and returned a few steps further on?"

Witness

Testaments of Sand

The sediments are a sort of epic poem of the Earth. When we
are wise enough, perhaps we can read in them all of past history.

Rachel Carson, *The Sea Around Us*

BURIAL

The girl sat on the sand by the river, sifting the grains over and over through her
fingers, as three-year-olds tend to do. Her attention was divided. First, she needed
to keep track of her mother, who was moving down the riverbank through the sparse
trees, picking what fruit she could find. But upstream, above the hills, something
fascinating was happening. A great, dark, turbulent mass of clouds had been rolling
over the landscape, shot through with lightning and shaking with thunder. The sky
had turned a strange earthy color in the late afternoon sun, and the ash cloud from
the volcano on the horizon added its brushstrokes to a scene that transfixed the little girl. A distant rumbling sound began and rapidly grew louder. The girl turned
and saw her mother running toward her. But it was too late: out of the river valley roared the first wave of a flash flood, tumbling the girl like a ball, forcing her
into the bank, where the torrent of sand and pebbles quickly covered her.

Over three million years later, Zeresenay Alemseged picks away the sand grains
from the girl's skeleton, one by one; he has been at the task for years. Dr. Zeresenay
is a paleoanthropologist, working at the Max Planck Institute for Evolutionary
Anthropology in Leipzig, Germany, far away from his home in Ethiopia. But it was
Ethiopia that yielded to him his treasure, the skeleton of the toddler in her tomb
of sand. She had lived her short life in what is today the Dikika region of the Afar,
the searingly hot depression at the northern end of the East African rift valleys,

close to the Red Sea. The rift valleys have long marked, as their name implies, the rent in the Earth's crust along which Africa threatened to split apart; in the Afar Depression and the adjacent Red Sea, this rip is active today, as evidenced by volcanoes and faults that create new land every year. But in occasionally more benign times, this was the home of our ancestors. It's where the celebrity fossil Lucy was found, but the "Dikika baby," remarkably preserved, is perhaps on her way to taking over the star role. Lucy and the child are of the same species, *Australopithecus afarensis,* but the child lived much earlier than Lucy. They are representatives of a crucial stage in our evolution, and the Dikika child, tiny though she is and still not fully revealed, has already shed light on critical questions. The sand, in burying her instantly but not violently, has preserved incredible details of her small body. Her face, braincase, lower jaw, all but two of the teeth (including adult teeth still waiting), collarbones, vertebrae, ribs, fingers, and kneecaps have all been revealed so far. Zeresenay has found, in her throat, the minute hyoid bone that is crucial for human speech. Though many of her features demonstrate that she had advanced only in some ways from her ape ancestors, she walked on two feet, and it is clear that the Dikika girl was on our evolutionary journey.

Sand preserves. Not only did it preserve the skeletons of Lucy and the Dikika girl, but it preserved *itself.* The layers of sediment in which our ancestors are found tell us about the rivers and the lakes, the volcanoes and the floods. Zeresenay works with colleagues from around the globe—geologists and experts on fossil animals and plants—to build up a picture of the world in which *Australopithecus afarensis* lived, a picture that is recorded in the character and contents of the sands and interpreted through the eyes of the geologists.

MAKING ROCKS

In the last few chapters, we have followed societies of sand grains on the move in today's world, journeys that take place over great distances and long periods of time. But however epic those journeys, in the end all sediment is deposited somewhere, and there it may remain. The sand that engulfed the Dikika baby was buried by the cargo of further flash floods. Sand accumulates, building up layer by layer, each layer the product of a separate depositional event—the waning of a river in

FIGURE 31. Turbidite landscapes of the South African Karoo. (Photo by author)

flood, an avalanche of sediment down a submarine canyon, a winter storm. If the circumstances are right, these layers, or beds, can build up over time into enormously thick piles of sediment that record long periods of the Earth's history. Look at the Grand Canyon, for example, or the Roraima of the Venezuelan tepuis, the foundation for the highest waterfall in the world. The accumulation of sediment layers is one of the Earth's great construction projects, ultimately building entire landscapes.

Figure 31 shows one such landscape, that of the Karoo in South Africa, where vast thicknesses of seemingly endless layers of sand-dominated sediment have built up from repeated turbidity currents hurtling down submarine slopes more than 250 million years ago. But if these were once sands and muds on the sea floor, how have they become solid rock, and how have they come to form an arid African landscape?

In the Karoo, loose, unconsolidated sand has been turned into sandstone, or *lithified*. Just as cooking turns cake batter into a cake (or something resembling a rock, if overdone), so lithification—a complex cooking process of temperature, pressure, and chemistry—turns sand into sandstone. The recipe for turning loose sand into solid rock reads roughly as follows: Pile up layers of sand successively so that the weight of the overlying sediment squeezes out the water from between the grains of the lower layers; allow the water to escape. Utilize the increasing temperatures below the Earth's surface to dissolve some of the grains in the remaining water, at the same time allowing minerals to precipitate and cement the grains together. The greater the mineral and chemical variety of the grains used, the more complex will be the result. Approximate cooking time: millions of years. As we saw in chapter 2, adding bacteria can speed up the job; as the head of the research project looking at *Bacillus pasteurii* at the University of California at Davis, has commented, "Starting from a sand pile, you turn it back into a sandstone."

But sediments themselves are only part of the story—after all, they can't simply keep on piling *up* unless the base of the pile is sinking. And wherever great thicknesses of sediment have accumulated, that's exactly what has happened: areas of the Earth's crust have subsided over long periods of time, constantly creating the space for further accumulation of sediment. Fill up a bathtub with sand and there's a limit to how much it can contain—unless somehow the bottom of the tub is sinking, in which case, as long as it continues to sink, more and more sand can be added. This is exactly what has been happening in the Afar Depression, preserving Lucy, the Dikika baby, and countless others of our ancestors. The Afar is called a *depression* for good reason—it is the lowest region of Africa, parts of it 150 meters (500 ft) below sea level, the result of the rifting of the land on either side. The Afar continues to sink, and sediments continue to pour in: it is a sedimentary *basin,* the geological term for our bathtub. Basins have been the great repositories, the storehouses of sediments throughout the Earth's history. They come in many shapes and sizes, depending on the causes of crustal subsidence. And there's feedback that amplifies the subsidence—the more sediments accumulate, the greater their weight, which further depresses the foundation of the basin. As long as subsidence continues and as long as there is a supply of sediment, then a basin will continue to grow. And the very fact that a basin is a depression almost guarantees a supply of sediment—the basin is a low point, surrounded by higher points, and in between, by definition, are slopes, down which sediment is inevitably transported.

Basins can form as plates collide and override each other, the crust sinking in front of a growing mountain belt; they can form where the crust is pulling apart, as in the Afar, and in California, where lateral movements cause subsidence (and uplift); and they can form in the middle of continents or at their edges. Two hundred million years ago, the Atlantic Ocean began to form, the old continent (as we shall see later in this chapter) breaking apart, as the Afar is today, eventually to the point where molten material from the Earth's mantle surged up through the rift, creating new, oceanic, crust. That process continues today at the Mid-Atlantic Ridge; the upwelling of the molten material is a source of heat and uplift and, as the continents drift apart away from the ridge, their broken edges subside continuously and over long periods of time. The subsidence creates space to accommodate the volumes of sediment pouring off the continent (not to mention the constant rain of organic debris), and a basin is formed. Our Susquehanna sand grain, carried out onto the continental shelf of the eastern United States, was following the journeys of countless of its predecessors that had accumulated since the ocean began to open. There are areas of this continental shelf where the sediments have built up to a thickness of *10 kilometers* (more than 6 mi). Basins are, indeed, truly massive sediment warehouses.

And when sand is buried to a depth of 10 kilometers, it experiences a huge weight of overburden and temperatures of up to 200°C (400°F), a recipe for change. Put the ingredients for your favorite cake together and they are stable at room temperature; put them in the oven and that stability is gone—the ingredients change their chemistry and the batter is turned into a cake. It is the same for sediments heated and squeezed in the depths of a basin. Minerals that had been dissolved in the water of the pores, the spaces in between the grains, precipitate and solidify. Different waters from deep in the sedimentary pile will flush through the pores, depositing new minerals. Vulnerable and unstable grains—for example, the feldspars that we saw weathered into clay in the first chapter—will decay and, again, change into clay and similar minerals. Even quartz grains, the survivors, will begin to dissolve, particularly along their edges, where they are crushed against one another. The net result? Grains are soldered together and the pore spaces between clogged up—the whole thing is glued together, cemented into solid rock, the finished cake of the Earth's kitchen (Figure 32). The chemical and physical—and sometimes biological—processes that turn sediment into rock are technically termed *diagenesis*. They are much studied but still far from well understood—rather like the mysterious behaviors of proteins and carbohydrates investigated by chefs who are experts in "molecular gastronomy."

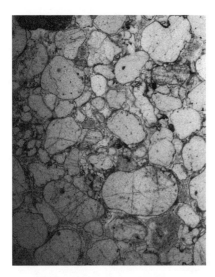 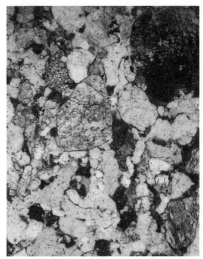

FIGURE 32. Very thin slices of sandstones viewed through the microscope. The darker material between the grains is the cement. (Left) A 280-million-year-old dune sand, almost entirely quartz grains, cemented by silica. (Right) Billion-year-old river sand, angular and poorly sorted; the large mottled grains are feldspar. (Photos courtesy of Tony Dickson, Cambridge University)

So, from sand a sandstone is formed in the bowels of the basin. And there it will stay unless the processes that have been causing the subsidence of the basin cease and, indeed, reverse themselves. The crust may rebound or the contents of the basin may be caught up in battles between the plates, wrenched and bulldozed upward, eroded, sculpted, and exposed for the benefit of inquiring geologists—and others. The ancient sands of Roraima and the Grand Canyon, our sand grain's relatives along the banks of the Susquehanna, and the endless landscapes of the Karoo have all experienced this history.

One key dimension of diagenesis is that the original sediment's character is preserved. It may have been altered and lithified, but the rock has not melted; many of the grains are intact, and the sedimentary patterns, texture, and architecture that were imprinted and constructed when the sand was deposited are, like the Dikika baby's bones, still there for all to see.

These great piles of sedimentary rock layers are ledgers: they record conditions of sediment accumulation and changes over long periods of time. As the last few chapters have described, on every scale from the individual grain to the basin, written in these ledgers is the planet's biography as told by individuals, tribes, and societies of sand grains. The stories are written in an arcane script, and it is the geologist's task to decipher it.

IDEAS

Stand in front of a pile of sandstone layers in the Karoo, or by a road cut, or next to a beach cliff: how do you read their stories, put together a mental image of their origins? The history of the deciphering of the language of sediments is a detective story and a soap opera of controversy, dogmatism, rivalry, and politics—a typical tale of the development of scientific ideas, with a colorful cast of characters.

Revolutionary ideas have often come from seemingly unlikely places. Nicolaus Steno, born in Copenhagen in 1638 as Niels Stensen, was a geologist for all of three years—in spite of the fact that the name Stensen means "son of rock." Yet his ideas were far ahead of their time, ideas that would contribute to the birth of modern geology 150 years later. Steno was from a wealthy family and trained in medicine, becoming physician to the grand duke Ferdinand from the Medici family of Tuscany. After dissecting a shark's head, at the duke's request, he realized that the teeth were identical to objects found entombed within rocks, at that time referred to as *glossopetrae,* "tongue stones," which were thought to have fallen from the sky or grown organically within the rock. Steno took the simple view that the objects looked like sharks' teeth because they *were* sharks' teeth, buried in the sand after death, and that the sand had now become dry land. Other contemporaries felt the same way (notably Robert Hooke in England, who had correctly declared earthquakes a cause of uplift of the land), but Steno took the whole matter several steps further, interpreting the layers of rock within which such fossils were found. In 1669, he published the ponderously titled work *De solido intra solidum naturaliter contento dissertationis prodromus* (roughly, *Preliminary Discourse to a Dissertation on a Solid Body Naturally Contained within a Solid*). In it, he described how the layers are formed by the settling of grains from water, that the older layers are over-

lain by younger, and that all layers are originally formed horizontally—any devia-
tion from this must reflect later events.

Herein lay the foundations of *stratigraphy*, the study of sequential sedimentary
layers to interpret the story of the time period over which those layers were deposited. The simple notion of *superposition*, that a sequence of layers represents the
passage of time, was radical—but it is fundamental to our reading of the ledgers
today. Unfortunately, even before this extraordinary work was published, Steno had
converted to Catholicism and found that this conflicted with his science, which he
abandoned, becoming a bishop later in life.

The same conflict that turned Steno away from science would inhibit development of his ideas until the end of the eighteenth century. Religious doctrine required a young Earth (born at 9:00 A.M. on October 26, 4004 B.C., according to
Archbishop Ussher) and held that all its features could be explained by the multi-
farious activities of the Great Flood.

Uniformitarianism: the challenge of saying this word belies its meaning, a simple but powerful idea about how to look at the Earth's processes. Uniformitarianism asserts that the things we can see going on today can account for everything
we see of the Earth's past: how continents and sand grains came to be, the landscapes and deserts of eons ago. It is often summed up simply as "the present is the
key to the past." Without this, any attempt to describe our planet would be nonsense: causes and processes could be explained by any harebrained, unrealistic, un-
necessarily complex, mystical, unobserved, and unrepeated mechanisms that might,
for whatever reason, appeal. Uniformitarianism works extremely well. The idea was
proposed in the eighteenth century by the British geologist James Hutton but was
dismissed by those who favored *catastrophism*, the shaping of the Earth by sudden,
cataclysmic events, such as the Great Flood; catastrophism was consistent with traditional ecclesiastical thinking. Hutton, like Steno, was fortunate to be born (in
1726) into a wealthy family, in his case landowners and merchants in Scotland. And,
like Steno, he studied medicine; the focus of his interest was, however, chemistry.
In 1747, he fathered an illegitimate child and, probably because of this, moved to
Paris to continue his studies. He eventually returned to Britain, took up farming
and chemistry, and developed an interest in geology—or "natural philosophy," as

FIGURE 33. Hutton's unconformity at Siccar Point. (Photo © Dr. Clifford E. Ford)

it was then called. He read Steno and Hooke and participated in the scientific society of the time, James Watt and Adam Smith being among his friends. His farming led him to observe soils and the underlying rock, and to put together ideas on erosion and deposition, none of which sat well with the doctrine of the time. He saw geological renewal and long cycles of events, all of which required the Earth to have a long history: as he remarked in a 1788 paper presented to the Royal Society of Edinburgh, he saw "no vestige of a beginning—no prospect of an end." The most fundamental of Hutton's observations was made during a boat trip with friends, including fellow geologist and mathematician John Playfair, along the coast of Scotland. At Siccar Point, a location that has become the destination of geological pilgrimages ever since, Hutton displayed the astonishing insight that was to change our understanding of our planet forever. Figure 33 shows these remarkable rocks, essentially the same today as when Hutton first saw them.

At their base are layers that are almost vertical, but these are truncated along a rough surface, above which are essentially horizontal layers of red sandstone. In a groundbreaking leap of the imagination, Hutton read the story of these rocks. The vertical layers are older, had once been horizontal, but had been tilted, uplifted, and eroded. The layers above are younger sediments deposited on the old eroded land surface. The boundary between the two groups of rocks must reflect a long period of time for this to have been accomplished—not thousands, but millions, of years. (This type of boundary is termed an *unconformity*, a break in the sequence of events recorded in the rocks.) We now know that the horizontal sandstones (of which we shall see more later) are tens of millions of years younger than the vertical layers below.

Hutton concluded that the ledgers could be interpreted by understanding the Earth's dynamic processes today and applying that understanding to the evidence of its past. The term *uniformitarianism* would not come into use until later, but the ground was prepared. In his Royal Society of Edinburgh paper, Hutton wrote:

> In examining things present, we have data with which to reason with regard to what has been; and, from what has actually been, we have data for concluding with regard to that which is to happen here after. Therefore, upon the supposition that the operations of nature are equable and steady, we find, in natural appearances, a means of concluding a certain portion of time to have necessarily elapsed, in the production of those events of which we see the effects.

Hutton's reasoning was revolutionary—and highly controversial, even heretical. He was attacked from all sides. His cause was not helped by the fact that his writing style was somewhat odd and opaque, often described as virtually unreadable (the quotation above displays unusual clarity). He was far better at defending his ideas in conversation; in spite of his manner being described as "peculiar," he was articulate, lively, and enthusiastic. Although his ideas would have to wait until after his death to be fully explained (by his friends, notably John Playfair), his intellect was remarkable, and his contribution to our ability to read the Earth's ledgers was radical and essential. Among those influenced by reading the unreadable Hutton's works was Charles Darwin.

During the nineteenth century, profound developments in knowledge and ideas about the Earth came thick and fast, always interwoven with soul searching over, and conflict with, religious doctrine. Those who found their first love to be geology included farmers, doctors, lawyers, and independently wealthy gentlemen; the stages on which they strode were three of the world's great mountain ranges: the Alps, the hills of Scotland and Wales, and the Appalachians. Among these geologists was Louis Agassiz, born in Switzerland and later the founder of the Museum of Comparative Zoology at Harvard University. One of the founding fathers of American science, Agassiz brought his knowledge of glaciers to the understanding of the ice ages and glacial landscapes and, in spite of being a firm opponent of Darwin's ideas, he became a major contributor to biology and ancient life studies. Charles Lyell, whom we met in chapter 5 observing the crumbling coasts of England, took Hutton's and Playfair's ideas on uniformitarianism many steps forward; only a few days before his death in 1875, he was working on revisions to the twelfth edition of his *Principles of Geology.* The list of the geologists of this period is long, their bewhiskered, serious faces staring out at us from the geological hall of fame.

Reading the script of sandstones, interpreting the ledgers, is intimately entwined with the entire science of geology. The nineteenth century saw the rules developed— rules that today seem blindingly obvious, but the intellectual insights that led to those rules were, in fact, extraordinary. Building on the work of Lyell and others, two men stand out during those years as founders of sedimentary geology and stratigraphy. Amanz Gressly, born in Switzerland in 1814, spent much of his life working in the Jura Mountains—indeed, today he is celebrated in his native land to a far greater extent than elsewhere. Trained, again, in medicine, he turned to geology at the University of Strasbourg. In the course of his work, he met Agassiz, who encouraged and promoted his ideas. Apart from having a dinosaur named after him (the obscure *Gresslyosaurus,* based on a few bits and pieces of bone), Gressly's legacy is one single publication—but it is an insightful and widely influential one. Fundamental to his thinking was the simple idea that if ancient sediments can be read correctly, then we should be able to make *paleogeographical* maps of the geography of the Earth's past, showing lagoons, tidal flats, beaches, dunes, and barrier islands, in the same way we make maps of the geography of today. To do so,

Gressly argued, we must recognize the characteristics of a sediment (and therefore a sedimentary rock) that indicate the environment in which it was deposited—river versus beach, deep water versus shallow, and so on. The principle of uniformitarianism is vital here—if we observe the shape and structure of a point bar today and see exactly the same character preserved in an ancient sandstone, then the simplest explanation is that the sand was originally deposited in a point bar; if we see ripples on the surface of a sandstone, then their geometry can be compared to ripples today to reveal the environment in which the sandstone was deposited. Gressly approached this in a way similar to that used for the recognition of facial character, defining associations of features with a particular *facies,* a specific depositional environment. Each facies is distinctive: the characteristics of sediments deposited by a meandering river are different from those of a braided one, and all of them are completely distinct from those of a barrier island. Aiding diagnosis, marine fossils will not be found in a river's floodplain.

He then reasoned that if you walk today from the landward side of, say, a barrier island, toward the ocean, you will cross a number of different but contemporary environments—lagoon, dune, beach, foreshore, and shallow marine sediments. These same kinds of lateral changes can be recognized by tracking changes in ancient sediments. And, critically, imagine that sea level rises, encroaching across the barrier island. If the sediments are preserved, then shallow marine sediments will be deposited on top of the earlier beach, beach sediments on the earlier dunes, and so on. The facies are stacked *vertically* in a sequence that is equivalent to the *lateral* sequence and record changes in depositional environment over time. These are among the most fundamental rules for interpreting the Earth's history that continue to be used today. There is, however, something of a mystery as to how Gressly derived these revolutionary ideas; after his single 1838 publication on these principles, he seems to have had nothing more to say about them, devoting the rest of his life, between bouts of mental and physical illness, to collecting and describing fossils.

Henry Clifton Sorby was an entirely different character, renowned as the greatest scientist that the steel town of Sheffield, England, has produced, and acknowledged as the father of sedimentology. Sorby was twelve when Gressly published his famous paper, and Ralph Bagnold was twelve when Sorby died. Sorby thus bridged nineteenth- and twentieth-century geology. His father owned a tool-manufacturing business that provided Sorby with financial independence. He was determined from the start to be a scientist, but since there were no appropriate courses, he never attended university. Instead, he simply set up a scientific work-shop and laboratory in his home. Sorby, in the spirit of Antony van Leeuwenhoek, was fascinated by the potential of the microscope and pioneered a method of grinding slices of rock so thinly that light could be shone through them and they could be examined microscopically (as in Figure 32). He understood, despite the derision of some of his fellows, that much could be learned about the large scale from careful study of the extremely small. In 1858, writing on the microscopic structure of crystals for the Geological Society of London, he remarked: "In those early days people laughed at me. They quoted Saussure [a giant of Alpine geology] who had said that it was not a proper thing to examine mountains with microscopes, and ridiculed my action in every way. Most luckily, I took no notice of them."

Sorby's work fascinated John Ruskin, the widely influential art and social critic, painter, and philosopher on the natural world. However, their friendship did nothing to alleviate Ruskin's agonizing over the aesthetic relationship between science and art (not to mention the century's ongoing conflict between science and religion). He was fascinated with the details of nature that Sorby revealed and at the same time dismissive. In a letter to his friend, the Pre-Raphaelite artist William Holman Hunt, he declared: "There's nothing makes me more furious than people's looking through microscopes instead of the eyes God gave them."

Sorby applied his microscopic innovations to metallurgy and biology, but he is best known for his interpretation of the features and relationships of mineral grains and what they tell us about sedimentary processes and ancient rocks and landscapes. He also initiated the quantitative analysis of sedimentary structures, recording the relationships between sand ripples, cross-bedding, current velocity, and grain sizes. In 1908, in the introduction to the summary of his life's work, "On the Application of Quantitative Methods to the Study of the Structure and History of Rocks," he wrote: "My object is to apply experimental physics to the study of rocks." The young Ralph Bagnold was ready to take Sorby's objective further. And the early years of the twentieth century saw rapidly developing and revolutionary ideas: Johan August Udden's obsession with sand (chapter 1) was bearing fruit; Alfred Wegener, a young Austrian meteorologist, was making his first expedition to Greenland and formulating his ideas on continental drift; the newly established U.S. Geological Survey was interpreting the libraries of the Earth's history revealed by the landscapes of the American West; and the age of the Earth was finally being established.

VESTIGES

James Hutton's inability to perceive any vestige of the Earth's beginning set in motion a century of anguished and often vitriolic debate. In 1851, in the thirty-sixth volume of his *Works,* Ruskin summed it up: "If only the Geologists would let me alone, I could do very well, but those dreadful Hammers! I hear the clink of them at the end of every cadence of the Bible verses." Charles Lyell did not place any limits on the age of the Earth, but by the end of the nineteenth century, that view would be refined. Charles Darwin, somewhat rashly but in the interests of making the point, reasoned that 300 million years would have been required to create the amount of erosion he observed in southern England. Darwin's estimate set off a new round of debate, with scientists queuing up to disagree with him on this (and other issues). The most influential and formidable of these was the great Scottish physicist William Thomson, perhaps better known as Lord Kelvin. Kelvin's reasoning was based on thermal conductivity: his view was that the Earth has continuously lost its initial heat and that to cool from its primordial molten state to its condition in modern times would have taken between 20 and 400 million years. He settled on an estimate of 100 million. Even though his assumptions—and many of his statements—could be best described as sweeping, Kelvin's reputation as a brilliant physicist carried a great deal of weight. Darwin was shaken by the criticism: he withdrew his estimate and, as recounted in Tony Hallam's review of the controversy, wrote to Lyell to warn him, "For heaven's sake take care of your fingers: to burn them, severely, as I have done, is very unpleasant."

In spite of well-founded geological reasoning that threw doubt on his figures, by 1897 Kelvin had cantankerously reduced his estimate to twenty-four million years. But Henri Becquerel's discovery in 1896 of radioactivity in uranium, and subsequent demonstrations by Pierre and Marie Curie and Ernest Rutherford that radiation produces heat (thus counteracting cooling of the Earth), put an end to the debate. In 1904, Rutherford was to give a lecture at the Royal Institution in London, part of which would deal with the implications of his work for estimating the age of the Earth. When he walked into the lecture theater, he saw to his dismay that Kelvin was there. As Hallam relates, he later wrote:

> To my relief, Kelvin fell fast asleep, but as I came to the important point, I saw the old bird sit up, open an eye and cock a baleful glance at me! Then a sudden inspiration came, and I said Lord Kelvin had limited the age of the earth, *provided no new source of heat was discovered.* That prophetic utterance refers to what we are now considering tonight, radium! Behold, the old boy beamed upon me.

It was now clear that the age of Earth should be measured in at least hundreds of millions of years. But exactly how old is our planet, and how can its age be measured? Rutherford set the direction by noting that the steady decay of radioactive elements could provide a clock. Early attempts to use that clock suggested that billions rather than millions of years would be the appropriate scale, but the methodology needed refinement. Committees for the British Association for the Advancement of Science and for the U.S. National Academy of Sciences in the 1920s and 1930s devoted a great deal of effort to this, and today, as we saw in chapter 1, the oldest direct evidence from Earth (from sand grains) points at an age of at least 4.4 billion years, while evidence from meteorites suggests an age of around 4.6 billion.

So, between the covers of our planet's ledger we have over four billion years' worth of records to interpret. How detailed and consistent are these records? Is the ledger complete? The answer to the second question is a resounding "no, nowhere near." Look again at Figure 33: Hutton's famous unconformity marks the position of tens of millions of years of missing records. But, of course, these missing records may have been preserved elsewhere; if we follow that break in the rocks far enough, perhaps some new layer of sand—a missing entry—will appear. Follow it further and what had been the break may now be represented by a huge thickness of layers—an entire missing account. But this reconstruction has to be painstakingly put together following the rules of our geological forebears. The process is like assembling a jigsaw puzzle, but given that a stack of sedimentary layers is the story of a period of geologic time, it is a four-dimensional puzzle.

FLOODS

Although the expanse of geologic time had not yet been fully defined before his death at the age of fifty in 1919, Joseph Barrell, a geologist at Yale University, published some remarkable ideas about reading the layers of the Earth's history. Barrell recognized the importance of knowing how much time is *not* represented in an apparently continuous pile of sedimentary layers, how many entries and records are missing from the ledger. Barrell, again oddly not considered among the conventional pantheon of geological heroes, spent time observing Appalachian rivers ancient and modern (he developed the explanation for the Susquehanna water gaps described in chapter 4) and recognized the key role of base level. Major periods of sediment deposition occur when sea level is rising, not falling, and local variations in base level will have the same effect. He described rhythmic and cyclical changes in climate and other environmental factors that determine whether sediment will be deposited or not in any particular setting—and proposed that by far the majority of the ledger is missing. Even in an apparently continuous stack of sedimentary layers, all perfectly parallel to one another, there are multitudes of what archaeologists refer to as *lacunae;* Barrell referred to them as *diastems.* Despite these gaps, Barrell used his own work and the results of early attempts at radioactive dating to quantify geologic time. He proposed that the great explosion of life on Earth, carefully documented in the rocks of the Cambrian period in Europe and the United States, occurred between 550 and 700 million years ago. Today's consensus is 590 million years.

Barrell's insights and research also laid to rest (more or less) some of the problems with uniformitarianism. Hutton's original ideas had been variously translated, interpreted, and corrupted into two broad schools of thought: first, a literal interpretation that natural processes have always proceeded in the same way and *at the same rates* as we see them operating today, and, second, that while the laws of nature have always remained the same and worked in a consistent manner, the *intensity* with which they have operated has varied considerably over time. The latter view accepts that, on the grand scale, there have been periods in our planet's history when, for example, volcanic activity has been more widespread than today, and the temperature of the atmosphere has been hotter or cooler; it also accepts that rates of erosion would have been different before plants colonized the land. As we have seen, human activities change the rates of sediment delivery to the oceans (the period of human influence on the planet has variously been referred to as the *psychozoic,* the *anthropocene,* and, more facetiously, the *mental*). On a small scale, every sedimentary layer represents a unique event, separated by Barrell's diastems, or periods of inactivity. Some sedimentary events record the periodic occurrence of great floods—but not *the* great flood.

The residents of New Orleans will vouch for the import of sudden events. Just as rivers burst their banks during times of flood, carrying vast volumes of sand out onto the floodplain, so did the London Avenue Canal during the storm surge of Hurricane Katrina on August 29, 2005. Flushing sediment out of Lake Pontchartrain, the surge burst the levee, and as it subsided, it left deposits of sand up to 1.8 meters (6 ft) deep in the city's streets and backyards; cars were buried in sand. Following in the footsteps of Sorby and Bagnold, geologists used the structure of the sand deposits—cross-bedding and so on—to reconstruct the direction of the current flow, its magnitude, and how porches, cars, and kitchen designs influenced deposition and erosion. This was not a depositional event consistent with a passive uniformitarianism; it was a catastrophic event. After all, deposition of sand occurs because the energy in a current, be it water or wind, is high enough to transport the sand, and as the energy level drops, so does the sand—but it's the high energy that starts the whole process.

The flood deposits of Hurricane Katrina are, understandably, not preserved, but others are. Today, there is gathering interest in what is lightheartedly referred to as *paleotempestology*—reconstructing the records of old storms and hurricanes by taking cores of the sedimentary layers in key environments around the coasts, particularly around the Gulf of Mexico and the U.S. eastern seaboard, and recording and dating the telltale layers of sand. While debates over hurricane frequency and climate change continue, these studies contribute real data over a period of thousands of years. Following Barrell's rhythms, there seem to be different cycles of hurricane activity, some over a scale of tens of years, some over much longer periods—some evidence suggests that catastrophic hurricanes pounded the Gulf Coast more frequently between 1,000 and 3,500 years ago than they do today.

Understanding the rhythms of destructive natural events is key to planning how to deal with them, whether floods, storms, hurricanes—or tsunamis. If you look at the "before" and "after" satellite images of the coasts devastated by the Indian Ocean tsunamis of December 2004, you will notice that beaches have often disappeared. Where has the sand gone? Inland. In the aftermath of the tsunami, vast areas of low-lying land along the shore were swathed in layers of sand, each layer scoured out from the seabed and the beaches and carried far inland by successive tsunami waves. The sand layers often show graded bedding, the coarsest grains settling out first as the pulse of water subsided. Tsunamis (literally "harbor waves" in Japanese, reflecting their destructive effects on ports) are very different from the normal ocean waves discussed in chapter 5. For a tsunami, the wave is not just the surface form of the water, with the water being less and less disturbed with depth; a tsunami moves the entire volume of water. Whether it's caused by the sudden faulting and shifting of the ocean floor, a giant submarine landslide, or a volcanic eruption, the ocean is displaced, like a bathtub full of water that has been tilted. And, in shallow coastal waters, there's nowhere for the water mass to go but upward, building into a towering wall that surges landward rather than breaking like a conventional wave. A tsunami has the power to carry and dump huge volumes of sediment far inland, covering tidal flats, fields, marshes, floodplains—a continuous sheet of sand where such a thing should not be, but sand that through its composition, thickness, sorting, grading, and other characteristics gives us vital information about how tsunamis work—and how, in the future, people can avoid the areas that they threaten.

Go to the flat lands around the mouth of the Salmon River on the central Oregon coast and dig a trench. Not far below the surface, after digging through the soils and muds typical of the place, you will come across a layer of sand. It may be only a few finger widths thick, but it clearly differs from the dark soils above and below it. If you look at it with a magnifying glass, it looks like beach sand and contains the shells of microscopic marine organisms. It's clearly out of place, not part of the normal sequence of events—it's a tsunami sand. In places, tsunami sand has been found directly on top of the remains of fire pits dug by the native inhabitants—perhaps abandoned in panic as the water surged toward them. Local stories and legends, like that of the raven with which this book began, talk intimately of the sea; but some are violent, recounting walls of water that rise up, flooding the land, destroying villages. The Pacific Northwest has a history of earthquakes and tsunamis because of its address—directly above the Cascadia subduction zone, where the Pacific crust is being forced down beneath North America (hence the volcanoes). After the earthquake and tsunami recorded around the Salmon River, the ground sank and the sand was quickly buried and preserved. This particular event has been recognized from British Columbia to Northern California and has been dated (in part thanks to evidence like the fire pits) to a period of twenty years around A.D. 1710. No historical records are available from North America, of course, but there are detailed accounts from Japan of a major tsunami on January 27, 1700. All the evidence, starting with the layer of sand, points to a Cascadian earthquake during the night of January 26, potentially as big as the Sumatran one of 2004. The resulting tsunami spread out across the Pacific, hitting Japan the following day. This was not the only tsunami to strike the Pacific Northwest in historical times—numerous other out-of-place layers of sand are building up the picture of frequent events over the past few thousand years.

While we associate tsunamis primarily with earthquakes beneath the sea floor, anything that displaces ocean water can cause one—sometimes in unlikely places. Along the east coast of Scotland is found a layer of sand that doesn't belong where it is, sandwiched within coastal peat. It contains the remains of marine organisms and has been dated to around 6000 B.C. It looks like a tsunami sand—but how could a tsunami have originated in the North Sea, which is far from any significant earthquake territory? The story came together when Norwegian geologists identified and dated a series of gigantic submarine rockslides on the floor of the Norwegian Sea. Around 6000 B.C., sections of the Norwegian continental shelf and slope failed catastrophically, sending debris hurtling into the deep water; when it settled, the mass of sediment covered an area the size of Scotland. Known as the Storegga slides, they were very likely the cause of the tsunami. Thus a thin layer of sand is detailed testament to a major natural event that occurred thousands of years ago, terrifying, no doubt, our Scottish ancestors, who were still dealing with the rising sea level following the retreat of the glaciers. The presence of wild cherries and the growth pattern of fish bones buried in the sand suggest that the tsunami struck in the autumn.

As we peer further back into the Earth's past, archaeology turns into geology. But there is a long period of time over which the two disciplines overlap, and today the light that can be shed on our own history and mythology by detailed geological investigation is being increasingly demonstrated—often by sand and its associates. Helen of Troy, whose face launched a thousand ships, would have gazed out at a very different scene—today, the ancient city of Troy lies far from the sea. But geological analysis of the sands and silts that clogged up the harbor and the estuary allows reconstruction of a geography consistent with Homer's accounts. Elsewhere in antiquity, the underwater sandbar that later grew to connect Tyre with the mainland probably enabled Alexander the Great to conquer the city. The development and decline of ports and settlements around the Mediterranean and elsewhere tell similar stories: the ill-fated harbor at Aigues-Mortes (chapter 5), Roman dams in North Africa that attempted to stop sand from destroying agricultural areas, and even, possibly, Atlantis.

Plato's accounts (hardly firsthand) put Atlantis in front of the straits called the Pillars of Heracles—an ancient name for the Straits of Gibraltar—and describe the island's destruction as coming from the Atlantic in the form of violent earthquakes and floods around twelve thousand years ago. Today, the Straits of Gibraltar mark one of the places where the tectonic struggle between Africa and Europe that formed the Alps continues. In 1755, Lisbon was all but destroyed by one of the largest earthquakes ever (it was felt as far away as Finland) and its aftermath. Tsunami sands around the Iberian coast record these events, and deep in the Gulf of Cádiz is a succession of turbidite deposits, each one bearing witness, as off the Grand Banks, to a major earthquake. These events have happened, on average, every couple of thousand years, and the largest of these deposits, potentially triggered by a very large earthquake, has been dated to around twelve thousand years ago. West of the Straits of Gibraltar is a large submarine bank, today 50 meters (160 ft) below the surface. Twelve thousand years ago, as sea level was rising after the ice age, the bank would have comprised a number of small islands. They wouldn't have been far enough above the surface to have been habitable; however, if the movements of regular great earthquakes caused subsidence, as they often do, could there once have been a larger area of land, an island of the size Plato describes—Atlantis?

EPISODES

The volume of sand and sandstone in and on the Earth's crust is enough to build the Great Wall of China around the equator fifty million times. And if every sand grain has a story to tell, that's a lot of stories—libraries of ledgers. For the remainder of this chapter we shall read, successively further into "deep time," a sampling of those stories, vignettes, of dramatic and workaday episodes of our planet's past to which sand bears witness.

The Mother of All Tsunamis

The dramas of modern earthquakes, tsunamis, and volcanic eruptions cannot compare to what happened on a uniquely bad day sixty-five million years ago. An extraterrestrial projectile around 10 kilometers (6 mi) across (although there are alarmingly smaller estimates), traveling at perhaps twenty times the speed of sound, slammed into what is now the Gulf of Mexico. As catastrophes go, this one is virtually indescribable. In the blast of the impact, molten rock, dust, and noxious chemicals were flung into the atmosphere, where they remained for hundreds of thousands of years. The hole excavated instantaneously by the impact is more than 180 kilometers (110 mi) wide; this crater is now buried beneath younger sediments of the Gulf and the Yucatán Peninsula, where the town of Chicxulub has, unpronounceably, given its name to the feature. The catastrophe is notorious for causing the extinction of the dinosaurs—together with upward of three out of every four species on the planet. There are, as usual, debates about this. The extinction was not instantaneous, but rather a slow death over several hundred thousand years, and odd creatures—crocodiles, for example—survived. There is even some evidence that one or two small dinosaurs made it. But given that the atmosphere was poisoned, the light of the sun shut out, the land covered in fires, the rain acid, and the air choked with dust, it is hardly surprising that these were difficult and complex times. The details are still being worked out, particularly with the aid of crater data gathered through drill holes, and one of the challenges is that the immediate area of the impact retains very little evidence of the event itself, since the impact effectively removed its own evidence, creating a gap, a diastem, lasting hundreds of thousands of years.

Away from the crater, however, there is ample and dramatic evidence—as well as numerous unconformities. The impact created the mother of all tsunamis, as did the water rushing back into the hole, and some extraordinary deposits of sand resulted—extraordinary simply because of the material and the velocity at which the currents were moving. Researchers from Japan, well versed in the art of modeling tsunamis, have made some rough calculations of the waves, suggesting heights of 200 meters (650 ft). These horrifically colossal waves hurtled great distances, leaving death, destruction, and deposition in their wake. Evidence of both the depositional and erosional effects of these waves is found all around the Gulf of Mexico and the Caribbean, across the Mississippi Valley, along the Atlantic coast of the United States, and possibly at locations in South America and Europe. The composition and structure of sand layers around the Gulf Coast record reversals of current direction as the waves came and went; the grain size variations tell the story of velocities and frequencies, and unusual cross-bedding tells of extremely high and rapid flow. Gravity slides and turbidites record the collapse of submarine slopes. And most of the sands contain strange glassy grains known as *spherules* or *microtektites*. The impact not only melted rocks, it vaporized them, ejecting a fireball of material into the atmosphere. There, the vapor (and the molten material) condensed and fell back to Earth as a rain of minute glass spherules. These are now found as grains of sand, but grains with a very unusual chemical and mineral composition; analysis of their chemistry indicates that they were formed at temperatures of more than 1,500°C (2,600°F).

The sand deposits and grains tell extraordinary stories of a (fortunately) extraordinary event. However, as in all good detective stories, the evidence is sometimes conflicting: layers of sand contain spherules in the wrong part of the sequence, and there are other possible origins for the sand deposits. If all the evidence were straightforward, the story would be less exciting.

Dunes, Dilatancy, and Dinosaurs

Even allowing for what awaited them sixty-five million years ago, the dinosaurs had a good run, and they live on dramatically in the popular imagination. But for a long time, they were known only from fragments of skeletons, bits and pieces (like *Gresslyosaurus*), often imaginatively put back together. Then, in the 1920s, expeditions from the American Museum of Natural History discovered a treasure trove in the Gobi Desert, a beautifully preserved abundance of what the ancient Chinese called "dragon bones"; even nests with eggs were buried, essentially intact, in the sand. Today, the museum and the Mongolian Academy of Sciences, together with other research organizations, cooperate to reveal this treasure. In particular, the site at Ukhaa Tolgod, the Brown Hills, is a gold mine of paleontological riches. Ten million years before the catastrophe on the other side of the world, dozens of species of mammals and reptiles enjoyed a good place to live, and, from a paleontologist's point of view, it was also a good place for them to die, for their remains are exquisitely preserved. Tiny mammals and dinosaurs sitting on their eggs have been painstakingly removed from the sand. But what kind of sand, and how could it achieve this extraordinary preservation? The creatures, like the Dikika baby in Africa, must have been rapidly buried, allowing no time for predators to dismember the bodies. On the face of it, the rich red sandstones in which the fossils are entombed bear all the hallmarks of desert dunes, ancestors of the Gobi sand seas of today. But today (with the possible exception of Cambyses and his army—see chapter 6), creatures are not suddenly buried by sand dunes—they move too slowly and predictably, and even dinosaurs couldn't have been stupid enough to simply stand around, waiting to be buried by a dune. Besides, how could dinosaurs have thrived if the place was like the depths of the Sahara today?

A clue came from the type of sand in which all the fossils were found. Associated with the fossil-rich layers were great thicknesses of sand that showed the characteristic cross-bedding of dunes (and, occasionally, dinosaur footprints), but the sand that entombed the dinosaurs had no such features—it was, in fact, featureless, a simple, structureless mass. Layers of mud between the dunes attested to the very different climate of the time, warmer and wetter, not unlike that of the Sand Hills of Nebraska today, where periodic torrential rains saturate the dunes and cause massive flows of waterlogged sand, which have been known to fill up buildings located in the shelter of a dune. In Mongolia, sandstones of the same age as at Ukhaa Tolgod show the remains of burrows made by creatures, like those today, escaping the heat of the day, but there are signs that they had to excavate new burrows as the old ones were plugged up with mud from the rains. At Ukhaa Tolgod, layers of sand were found that were cemented by a kind of calcium carbonate found in arid environments and known as *caliche*. This would have effectively blocked water from draining away, out of the dunes. Combine these pieces of evidence and perhaps the mystery is solved. Cloudbursts were probably more frequent at Ukhaa Tolgod then than they are in Nebraska today, and the dunes were significantly bigger. Saturate a towering dune face with water, prevent the water from draining away, destabilize the sand through the effects of dilatancy, and the slightest tremor, perhaps the wind, perhaps an irritated dinosaur, would cause an instantaneous slide of huge volumes of sand slurry—burying the irritated dinosaur. Here are desert processes, diagenesis, the physics of granular materials seventy-five million years ago, and forensics at work.

Breaking Up

The dinosaurs of the Gobi had cousins all over the world. In Philadelphia in 1787, Benjamin Franklin was presiding over a meeting of the American Philosophical Society. George Washington was also at the meeting, and together they examined a large, heavy bone presented by Caspar Wistar, a local physician, who had dug it up in New Jersey. It was identified as the thigh bone of an unusually large human. In 1802, the young and gloriously named Pliny Moody, working on his father's farm in Massachusetts, plowed up a slab of red sandstone on which was imprinted a series of large, three-toed footprints. They were decreed by religious authorities to have been made by "Noah's raven." Had either of these finds been correctly identified, they would have qualified as the earliest recognized dinosaur fossils, but that honor went to a discovery by the Reverend William Buckland in England some decades afterward. Although the term *dinosaur* would not be coined until 1842, Buckland, who consorted with Lyell and Agassiz and trod thoughtfully the difficult path between the Bible and the new geology, described to the London Geological Society in 1824 his finding near Oxford of the "*Megalosaurus* or Great Fossil Lizard of Stonesfield."

Dinosaurs were clearly as at home in New England as in Mongolia, and the eastern United States continues to yield a trove of bones and footprints, not of the very first dinosaurs (that distinction belongs to South America), but of some of the earliest. The red sandstones in which Moody discovered the footprints are found up

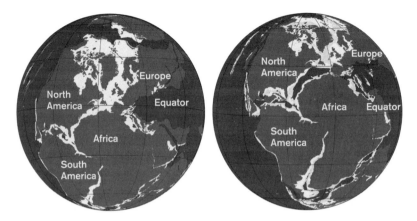

FIGURE 34. Pangea assembled, 260 million years ago (left), and fragmenting, 160 million years ago (right). The darkest areas show ancient oceans, and the light areas between the shapes of today's continents show their true structural extent, the continental shelves. (Images generated by Cambridge Paleomap Services Limited, modified by author)

and down the northeastern United States, and they record a major episode in the Earth's ledger: the fragmentation of a continent. The dinosaurs first appeared during the chapter of geological history called the Triassic, a chapter that lasted from around 250 million to 200 million years ago and whose beginning followed, as we shall see, a dramatic ending. By the beginning of the Triassic, most of the planet's continents had been welded together into one vast landmass, Pangea—or Pangaea, depending on which part of the now fractured supercontinent you have ended up on (Figure 34).

Two hundred and sixty million years ago, the northeastern United States lay just south of the equator, the United Kingdom and much of Europe just north. As Pangea drifted slowly northward, the western United States dwelt for a while in the same latitude as today's Sahara: go to Zion National Park and look at the Navajo Sandstone from this time, with its gigantic sweeping cross-bedding, and you will see evidence of ancient sand dunes. The architecture of the cross-bedding helps us reconstruct continental movement through different wind systems at different latitudes: although researchers at the University of Nebraska (who also work on the

Gobi dinosaurs and the Sand Hills) have demonstrated that the story, as is so often the case, is not simple, these ancient dunes have tales to tell us about the wandering supercontinent.

This arrangement of the Earth's landmass was not to last, however. Convection and churning of the Earth's molten interior would begin to break up the supercontinent; as if the brittle crust on the top of a crème brulée had been shattered, fragmentation began. It was not an overnight process—it took close to 100 million years for what we now know as parts of the Atlantic Ocean to form, new oceanic crust solidifying from the molten upwelling between the fragments. The process continues today.

Pangea broke up slowly and far from simply, not along a single crack, but through a mosaic of intersecting fracture systems, some of which would continue to open, others of which would stall. The network propagated over time, like a system of gigantic zippers, Africa pulling away from North America first, Europe from Greenland, and southern Africa from South America much later. The process began with faults, fractures in the Earth's crust allowing the crust to pull slowly apart. The result was a series of rift valleys, bordered by faults whose movement successively dropped the valley floors—think of the topography of the Great Basin or the Rio Grande today, or East Africa, the Red Sea, and the Gulf of Suez. Rift valleys were where our human history began—Lucy and the Dikika baby were both found in the great East Africa Rift system. These are among the most dynamic of tectonic and sedimentary environments, with towering valley sides along the faults and broad valley floors; movement on the faults constantly shifts the local base levels, and erosion of the bordering highlands constantly pours sediment into the valleys. It was across this kind of landscape that Moody's dinosaur strode.

These great Triassic rift valleys spread not only along the northeastern edge of North America, but across the United Kingdom and into Europe, and northward past Greenland (Figure 35). Although the continents were drifting northward over this period, for much of the time the rift valleys were forming under equatorial and desert climates. Tropical weathering, with iron minerals coating the grains and being concentrated during later diagenesis, resulted in almost all the sediments deposited in the rift valleys being red. For a long time, this group of rocks was known as the New Red Sandstone (and yes, there is an Old Red Sandstone, but more of

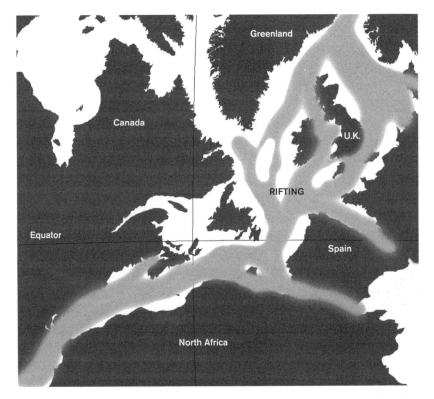

FIGURE 35. The rifting pattern that led to the breakup of Pangea. (Image generated by Cambridge Paleomap Services Limited, modified by author)

that later). I grew up in Nottingham, in the English Midlands, to which visitors flock in search of the vestiges of Robin Hood. What they find is a (relatively modern) castle built on a craggy outcrop of red sandstone of the Triassic period. The sandstone is easily excavated, and Nottingham's foundations are riddled with an entire world of tunnels and rooms: a tannery, a brewery, homes, and storage spaces. The oldest pub in England was partly excavated into the red rocks, and countless wells produced freshwater from the sandstone. Look at the walls of these excavations and you will see pebble beds, cross-bedding, muddy layers, and other diagnostic features that tell a story of lake beds, flash floods, and river sandbars.

In the countryside around Newark, New Jersey, and Amherst, Massachusetts, you will find the same features. Travel around Nevada or the Gulf of Suez today and you will see the modern setting for these deposits: steep valley walls, along which cascades of erosional debris build up great fans of sediment; rivers, dried up for much of the year but rushing torrents after rain; lakes that dry up to leave pale surfaces of salt and cracked mud; sand dunes; and, occasionally, the sea encroaching as the land level shifts. The terrain is shaken by earthquakes, which change the elevation of the valley floors, divert drainage, and empty or fill lakes—constantly shifting base level. Rift valleys are complex environments and the sediment record extends deep below the surface; Triassic rifts can contain 2,500 meters (8,000 ft) of sediments successively deposited as the valley floor foundered. Detailed work on these sediments on every scale—the reconstruction of sedimentary environments and input over space and time, and the directions of sediment transport—sheds light on the movement of faults and the development of the rifts. It tells the story of the breakup of a megacontinent.

These rocks have also had their say in modern history. Triassic sandstones built New York's brownstones. And the rifting of the crust allowed molten rock from below to find its way to, or close to, the surface—volcanic rocks are often associated with the red sediments. One such rift valley ran through Gettysburg, Pennsylvania, and the resulting topography played a critical role in the Civil War battle. The volcanic rocks, hard and durable, formed the higher ground occupied by the Union forces: Cemetery Ridge, Culp's Hill, Little and Big Round Tops. The Confederate troops moved through the lower ground, underlain by the more easily eroded sediments. The volcanic outcrops and boulders provided command of the battlefield and some cover but prevented digging in—the Union casualties were high. Nonetheless, the Triassic rocks, which mark the breakup of a continent, helped prevent the breakup of a nation.

The Mother of All Extinctions

As noted earlier, the ledger's entries predating the Triassic ended with a cataclysm. The final entry was the Permian, a climatically difficult period for life—but life was nevertheless thriving, both in sea and on land. Then, in a series of pulses of mortality or in one geologically sudden event (the evidence and schools of thought vary), 90 percent of marine species and 70 percent of land life were obliterated. Whole families and genera of creatures disappeared forever.

As an undergraduate student over forty years ago, I stood on an outcrop in the Arctic, one foot on the Permian, one on the Triassic. I admit that I did not fully appreciate the drama represented between my feet—perhaps because much of the science of mass extinctions had not yet been done. Beneath my Permian foot was ample evidence of healthy marine life—large corals and other fossils—beneath my Triassic foot, nothing but lifeless sandstone and shale. Corals were one of the marine groups hit hard at the end of the Permian, and those beneath my feet were never seen again.

The turbidite sandstones of the Karoo (Figure 31) are Permian in age. Toward the top of this sequence of sediments, the character of the marine environment disappears, reflecting a drop in sea level, and the sediments become typical of those deposited in rivers. Careful study of these river systems sheds important light on what was going on at the end of the Permian. Immediately beneath the Permian-Triassic boundary are sandstones and shales with all the characteristics of channels and floodplains of a gently meandering river system (as in Figure 21). But above, the sandstones are thicker, coarser-grained, and characterized by very different internal structures: the rivers are now braided (as in Figure 22). Braided rivers today form when the sediment supply is abundant, clogging up channels, building up channel sandbars, forcing the river to change course and break up into multiple channels. The extinction had demolished much of the plant life on the land—trees, bushes, and grasses—leaving the forces of erosion free to sweep across the exposed surface, generating a huge increase in the supply of sediment to the river systems. A similar change was seen in local river patterns after the eruption of Mount St. Helens.

The "smoking gun" for this devastating extinction is nowhere near as clear as for the event that wiped out the dinosaurs. There is chemical evidence of major climatic change, dramatic lowering of ocean oxygen levels, acid rain, and increase in atmospheric carbon dioxide. In Siberia, one of the world-record series of volcanic eruptions was going on around this time, and an extraterrestrial impact has been identified in Australia that may or may not prove to be the culprit. But here again, some fascinating details are being provided by microscopic forensics, this time from oil. Oil is formed by the degradation of organic material—microbes, plants, and animals—and these donate bits and pieces of their molecular identities to the oil. Even if it is only found as minute droplets preserved in microscopic cracks in sand grains, the oil reveals those molecular identities (known as *biomarkers*), which often tell us far more about the living world of the past than do conventional fossils. This kind of biochemical forensic work has suggested a cause for the great extinction at the end of the Permian that has nothing to do with an impact, but everything to do with a catastrophic change to the biological balance of the planet. The Permian oceans seem to have teemed with life, and the biomarkers indicate waters with an ample supply of oxygen. This benign environment was then abruptly terminated: oxygen levels plummeted, and the oceans were taken over by bacteria that can't tolerate oxygen but thrive on hydrogen sulfide, the poisonous gas with the smell of rotten eggs. The oceans had become stagnant and toxic, breathing their poisons into the atmosphere with lethal consequences. Did the Siberian eruptions create a runaway greenhouse world with temperatures rising so rapidly that the atmosphere could no longer oxygenate the oceans? And were many of the complex creatures that had evolved defeated, wiped out by primitive microbes? The forensics are still inconclusive, but the identities of the victims are well established.

Mountains Appearing and Disappearing

Where did the monumental volumes of sand that swept across the devastated post-Permian Earth and filled the rift basins of the Triassic come from? They came from the erosion of mountains—where most sand originates. The supercontinent of Pangea had been welded together by the closing of an earlier great ocean that for hundreds of millions of years had been the predecessor of the modern Atlantic. Like the western Pacific today, this ocean's story involved a complicated sequence of moving plates, large and small, volcanic island chains migrating as small ocean basins behind them were consumed by subduction. Fragments of continents and volcanic islands collided with one another and with the major continents over time, each of these collisions forming a mountain belt along the edges of the ocean. The Appalachians were the result of some of these crustal fender benders and pileups. And the Appalachians are part of a continuous chain of mountains from the Ouachitas in the American South through Newfoundland, Ireland, Wales, Scotland, and Norway, and reaching north into Greenland and Spitsbergen (where I stood astride the Permian extinction). This mountain range, collectively known as the Caledonides, after the Roman name for Scotland, was a bustling laboratory for the early days of geology, the source of great discoveries and great controversies, and continues to be today.

The Appalachians were constructed by three major pileups, 500, 400, and 300 million years ago, the last event completing the welding of Pangea. Each one crumpled the edge of the continent and bulldozed indiscriminately a miscellany of geological material up onto that edge—oceanic crust, volcanic islands, bits of other continents, older sediments. All of this, as newly upthrust mountains, began to crumble and erode. Indeed, some of the erosion was happening while the bulldozing continued, and piles of sedimentary detritus were cannibalized into the growing heap. The stories of mountain building, *orogeny*, whether it be in the Caledonides, the Alps, the Urals, or the Himalayas, are about the most complicated of any of the planet's history. How to unravel them, when most of the evidence has been removed? The Appalachians today are, after all, mere stumps, worn-down remnants of their former glory. Erosion may proceed slowly, but the Appalachians today are being worn away overall at an average rate of 3 millimeters per century—that's 30 meters (100 ft) every million years. On a geological scale, that's a lot of rock removed. (Interestingly, the valleys may be eroding faster than the summits, making the Appalachians increasingly rugged.) Young mountain ranges, where the variation in elevation is greater, show rates of erosion perhaps ten times that. It's in all that sediment, much of which is sand, that the opportunity lies for us to reconstruct the story of the mountain range, to put back together the sequential stripping away of the pile of bulldozed rock, rather like visualizing a tree from the sawdust around its stump.

This is where *provenance* really enters the game. Like a family genealogy or DNA forensics, the makeup of a family of sand grains preserved in a sandstone tells the story of its provenance, or where the family came from. Just as sand grains can be made of many things, so too can sandstones, but by far the majority of sandstones are made up of quartz plus a few other key ingredients. For a long time, sands and sandstones have been given names, or classified, on the basis of three key components: quartz, feldspar, and lithic (rock fragment) grains. Some sandstones that have endured multiple cycles through the mill of weathering and erosion have lost all components except durable quartz. Other, less mature, sandstones still retain grains of feldspar and rock fragments, which, in the longer term, are less stable and will disintegrate with time. Take a large sample of sandstones from the same sequence of beds; make, as Sorby did, very thin slices of them; peer down a microscope and count the grains, thousands of them, of each type; and finally plot the proportions of these types on a triangular diagram, with each of the three major components occupying an apex. (This sort of diagram is known as a QFL—quartz, feldspar, lithic—plot.) Different sandstones will cluster in different areas of the triangle, and different areas of the triangle are given different names—thus you will have classified your sandstone. You might have identified it as a *sublitharenite* or a *lithic subarkose,* arcane names that will not detain us here but are loved by sedimentologists. What is important is the principle of separating out different kinds of sandstone according to their families of grains, which in turn reflect their family tree.

As the theory of plate tectonics took hold, its power to link a particular tectonic setting (such as a volcanic island chain above a subduction zone, a rift basin, or the passive margin of a continent moving away from a spreading ocean ridge) with the associated sediments became clear. In the early 1970s, a geologist at Stanford University, Bill Dickinson, took this idea, did the work, and set out the rules. He took the traditional QFL diagram and defined areas within it that circumscribed sandstones derived from different plate-tectonic settings. The broad groupings were sand originating in old stable continents; sand recycled out of mountain belts; and sand deriving from volcanic islands. Over the years, the scheme has been refined and new approaches of mineralogy and chemistry brought into the equation— the science of provenance has been honed into a fine art. Dickinson went on to become one of the tectonic and sedimentological heroes of his generation. He recently published a lengthy and exquisitely detailed study of the provenance of sand grains included in prehistoric pottery (temper sands) of the Pacific Islands. Dickinson's intimate knowledge of the geology of the Pacific and his meticulous methodology documented patterns of migration, settlement, and commerce that had hitherto been impossible to define—all through the study of sand grains.

But back to the Appalachians. Provenance studies, combined with measurement and mapping of sandstone sequences, facies, and structures, make it possible to interpret the story of the sequential erosional stripping of a newly formed mountain range and the sedimentary environments in which the sediments were deposited. The three major episodes of the formation of the Appalachians each resulted in the accumulation of vast thicknesses of sediment on the flanks of the uplift—the very weight of the bulldozed pile depresses the crust and a deep basin forms in front of the advancing pile, a ready receptacle for the cascades of sediment. One of these sediment piles in particular qualifies as a famous sandstone, a well-studied section of the ledger: the Old Red.

The Old Red Sandstone—named to distinguish it from the New Red Sandstone— records over sixty million years of history (from the end of the chapter known as the Silurian period through much of the subsequent Devonian), over a huge stretch of territory from today's Arctic to the Gulf of Mexico. Such a scale of time and distance necessarily covers a considerable variety of geologically newsworthy events—there was simply so much going on at different times in different places. The opening and closing of small ocean basins and the consequent tectonic impacts in what is now Europe differed in timing and effect from what was happening on the Appalachian side of the story. A collision was not always head-on, but sometimes more of a glancing blow, tearing parts of the continent sideways (rather like California today), along laterally moving faults that ripped open deep depressions, into which sand poured. Seas invaded and retreated at different times in different places. Pieces of crustal flotsam drifted around the ancient ocean, beaching, colliding, piling up. It was a period of ponderous tectonic chaos that completely changed the geography of the world.

The Old Red represents a wide variety of sediment genealogies, and for this reason the term is used fondly but only informally today—these gigantic piles of sediment have been subdivided into more manageable and locally meaningful chunks. But the Old Red continues to have a place in geologists' hearts because of its prominent and dramatic role in the growth of geology as a science. The sandstone above Hutton's unconformity (Figure 33) is the Old Red. Agassiz discovered the remains of early fishes in the Old Red elsewhere in Scotland. And two members of the pantheon of nineteenth-century geologists, Roderick Impey Murchison and Adam Sedgwick, in 1839, used the spectacular sequences of the Old Red in Devon to define a new chapter in the Earth's history: the Devonian. Nevertheless, a great deal of effort was needed to establish the Old Red's importance. A colleague of Murchison's describes how a visiting foreign geologist told him, "You must inevitably give up the Old Red Sandstone: it is a mere local deposit, a doubtful accumulation huddled up in a corner, and has no type or representative abroad."

Murchison, thankfully, declined to follow this advice. The breadth of its stupidity can perhaps be portrayed by Plate 13. The spectacularly exposed sections of the Old Red Sandstone in Greenland record a thickness of over 6 kilometers (4 mi) of sediment accumulation. Even though the photograph was taken from an aircraft, cross-bedding can be distinguished in some of the layers.

The Devonian is an extraordinary record of an extraordinary period in the history of our planet. It is not surprising that it is, in itself, a textbook in sedimentology, displaying examples of almost every kind of sediment, every permutation of facies (not always red), from deep-water turbidites to sand dunes. As Richard Fortey wrote in *Life: An Unauthorised Biography*, "The two different faces of the Devonian—marine *versus* the deposits of lakes and mountain basins—are a kind of temporal schizophrenia. The non-marine rocks were lumped together as Old Red Sandstone, and it was some time before it was proved to everybody's satisfaction that these richly coloured red rocks recording life's greatest adventure were the exact contemporaries of unremarkable pale limestones and dark shales."

Here, in this passage from Fortey, is the other reason for the Old Red Sandstone's fame: "life's greatest adventure." For contained within the Old Red are the stories of the flowering of a sophisticated plant world and of some remarkable fish, together with tales of the rise of the vertebrates and their first expedition onto the land, recorded by their footprints in the sand. According to Mark Twain, in his 1903 essay "Was the World Made for Man?":

So the Old Silurian seas were opened up to breed the fish in, and at the same time the great work of building Old Red Sandstone mountains eighty thousand feet high to cold-storage their fossils in was begun. This latter was quite indispensable, for there would be no end of failures again, no end of extinctions—millions of them—and it would be cheaper and less trouble to can them in the rocks than keep tally of them in a book.

For a sampling of these great stacks of sediment that poured off each newly formed but crumbling mountain range, we shall return to the work of Barrell. In 1913, Barrell drew a map that showed a range of Devonian mountains extending from New York into Pennsylvania and, on their western side, what he called the Catskill Delta, in cross section a wedge-shaped apron of sediment flanking the mountains, extending and thinning (hence the "wedge") to the west and south. He proposed that the red color of many sediments told of an arid or semiarid climate and that the droughts resulting in such a climate were the impetus for animals to develop lungs. The sandstones along the Susquehanna that our sand grain hurtled past in chapter 4 belonged to the Catskill Delta. In places, the Catskill sediments are more than 3,000 meters (10,000 ft) thick. We now know that this accumulation of geological waste is not the result of a simple delta; rather, it represents a compendium of all the varied sedimentary environments that we might expect in front of a mighty range of mountains. The initial subsidence of the basin under the stress of the growing mountains was rapid. Early sediments in the sequence were turbidites, cascading into the basin, filling it up, flooding westward, until the sediments built up above sea level. The early river systems were braided, clogged with sediment, the later ones longer and meandering as the land built out and the sea retreated. Imagine taking the Appalachian Trail 380 million years ago. You are traveling considerably south of the equator, in a climate like that of Namibia's today. It's thirsty work, with little shade, no trees. Stop and look westward: in the foreground, steep valleys, rivers flowing out in wide braided channels onto broad floodplains; in the distance, meanders, sand banks, lakes glinting in the Devonian sun, in some places fringed by dunes. The scene shimmers in the heat. In the hazy far distance, the shoreline, the coast of the wide sea stretching far to the west, tidal flats, deltas, barrier islands, beaches—on which our ancient ancestors leave their footprints. The sea is slowly retreating westward, driven back, except for an occasional countersurge eastward, by the ever-growing land along the flanks of the mountains.

To take an Old Red Sandstone field trip today, stroll around the streets of New York, Washington, and other East Coast cities—your tour will reveal a diversity of facies (as will a tour of Scotland). The original curb and paving stones of New York are Old Red, and much of Yale University was built from the same stone—as was the spectacular but infamous Dakota Building, home of John Lennon and the scene of his death. Mark Twain, who seems to have had a great fondness for Old Red, wrote in an 1868 newspaper piece of "that poor, decrepit, bald-headed, played-out, antediluvian Old Red Sandstone formation which they call the Smithsonian Institute." Brownstones and curbstones, New Red and Old Red.

Devonian sandstones have been the subject of intensive geological scrutiny for more than two hundred years, and increasingly sophisticated methods continue to be brought to bear, for there remains much to learn. But through reading the stories told by their facies, grain sizes, provenance, chemistry, and all the other aspects of their character, the biography of a mountain range, its birth and death, has been told.

Time and Tides

If our ancient vertebrate ancestors on the Devonian shores had playful moods and, like Edward Lear's Owl and Pussycat, danced "on the edge of the sand . . . by the light of the moon," the Moon would have appeared to them to be bigger than it appears today. And the days they spent going about their ancestral business would have been shorter than they are today. The evidence is in the sands.

The most likely origin of the Moon is an almighty collision between the Earth and another planetary body shortly after the Earth was formed, the Moon spinning off into orbit, captured by the Earth's gravity. As water formed on Earth and the oceans gathered, they came under the tidal influence of the Moon—and the Sun. But we know that the Moon is slowly pulling away from us; direct measurements show this happening at a rate of 3.8 centimeters (1.5 in) every year. Celestial mechanics demonstrates how the never-ending cycle of tides puts a brake on the rotating Earth, energy is lost, and the Moon slips away. As the Earth slows down, the days become longer and there are fewer of them in the year; but regardless, the tide still comes and goes.

Tides, as we know, move sediment and thereby leave a record. The layers of sediment deposited by tides can be seen today (they are, not surprisingly, called *tidalites*), but they can also be seen preserved in very ancient rocks. In parts of South Africa, there are tidalites formed 3.2 billion years ago, when the planet was young and life did not add up to much yet. These rocks consist of multitudes of stacked repetitions of twinned layers, a fine sand deposited as the tide came in, and silt and mud as it went, more calmly, out. The layers vary systematically in thickness and in grain size, reflecting variations in the strength of the tidal currents, and these variations result from the tidal cycles that we see continuing today. Twice a month, the Sun and the Moon are lined up (the condition known, memorably, as *syzygy*) and conspire to produce a larger tidal pulse—the *spring* tides. Spring tides have nothing to do with the season, but refer to the leaping up of the tide. When the opposite configuration occurs and the Sun's pull detracts from that of the Moon, then the tides are subdued, the *neap* tides. A further cycle results from the fact that the Moon's orbit is not perfectly circular, and it moves closer and further away over the course of a month—the origin of the idea that *lunacy* is influenced by the proximity of the Moon, as in Othello's gloomy observation, "It is the very error of the moon, / She comes more near the earth than she was wont, / And makes men mad."

The cycles of the Moon influence the strength of tidal currents, which in turn determine the character of the tidalite layers. Measure the cycles of thickness and grain size in ancient tidalites, apply some mathematical statistical analysis, and the length of the day and of the lunar month can be estimated—along with how far away the Moon was. Three billion years ago, a month was around twenty days long, there were approximately 550 days in a year, and the Moon was probably 25 percent closer to the Earth than it is today.

Now, given how far geologists are peering back in time, with associated uncertainties over measurement, statistics, and interpretation, it is not surprising that this kind of conclusion is the subject of lively debate and dispute. But the approach continues to be refined, using tidal sands from different ages around the world—and it remains the only insight we have into the early relations between ourselves and our nearest celestial neighbor.

I suspect that James Hutton would have been delighted by tidalites and their stories. Interpreting the incomplete ledgers of our planet's history takes, as I hope this chapter has shown, hard work and imagination. In the words of John Playfair, describing in 1803 his friend Hutton's earlier reading of Siccar Point: "The mind seemed to grow giddy by looking so far into the abyss of time; and while we listened with earnestness and admiration to the philosopher who was unfolding to us the order and series of these wonderful events, we became sensible of how much further reason may sometimes go than imagination may venture to follow."

To see a world in a grain of sand.

Sand and Imagination II

Stories, Medium, and Muse

> One picture is worth ten thousand words.
>
> From an advertisement on a 1920s San Francisco streetcar

> One part of drawing in the sand that's really great is that,
> no matter what I do, no matter how big it is, I have a
> completely clean sheet of paper, meaning a completely
> clean strip of sand that I can return to every day, and
> there's an incredible freedom in that kind of artwork.
>
> Jim Denevan, land artist

WRITTEN IN SAND

Milpatjunanyi: it's a word that does not roll easily off Western tongues but carries that promise, common to ancient and exotic languages, of insight and discovery. All languages offer novelty and oddly satisfying demonstrations of the limitations of our own—multiple words to describe what in our language is a single color, for example—and the Pitjantjatjara language of the communities of north central Australia is no exception. Pitjantjatjara is one of the many languages of the many Aboriginal communities in the region around Uluru (Ayers Rock), a natural monument of ancient sandstone. *Milpatjunanyi,* in the Pitjantjatjara language, means the art of *telling stories in the sand.* It is a heritage deeply rooted in the land, in journeys, maps, and the dreamtime, a culture of stories whose narratives are actively interwoven with drawing in the sand. The storyteller, often a woman, prepares the canvas by sweeping her arm across the sand; the drawing is generally done with a stick, initially pressed against the narrator to create a ritual connection with the Earth, and the stick is often used as a drumstick to provide accompanying percussion to the story. Men and women have different rituals, different stories, but much of the skill is handed down by the women, using sand drawing as a teaching tool, in exactly the same way that wooden sand trays were used well into the nineteenth century for teaching British children to write. The Aboriginal stories and the accompanying sand icons provide the means of sharing knowledge and tradition within families and between distant communities.

Throughout the world and across cultures, sand has long provided—and continues to provide—the medium for drawing, writing, calculating, teaching, and divining; it is a medium for narrative art and, as we have seen with sand sculpture, art for art's sake. It is a readily accessible medium for which a finger is the only necessary tool; it is cooperative, yielding, tolerant, reworkable—and ephemeral. Although there are examples where it is trapped into solidity, the intrinsic character, value, and spiritual appeal of sand as a medium is its fragility, its impermanence. And what is startling is not simply the ubiquity of traditions of sand as a medium, but the common threads among the *ways* in which it is used, the designs and patterns. It seems almost to be part of our collective subconscious.

PATTERNS IN THE SAND

More than 3,000 kilometers (2,000 mi) east of Uluru lie the Pacific islands of Vanuatu, where the tradition of sand drawing has been proclaimed a "Masterpiece of the Oral and Intangible Heritage of Humanity" by UNESCO. These age-old visual designs transcend the differences among the eighty languages of the islands. They provide a rich means of communicating and teaching history, rituals, farming methods, family histories, myths, and, indeed, knowledge and stories of all kinds.

The Vanuatu sand drawings comprise a vast collection of intricate designs and patterns, often highly geometrical and made up of interwoven straight lines, curves, and loops—all drawn in a single continuous motion of the finger in the sand (Figure 36). The patterns are ancient and carry layers of meaning; they are mnemonics and records, story illustrations and choreography. As in the traditions of *milpatjunanyi,* women play a leading role in teaching the art and its forms.

From Vanuatu, travel westward, back across Australia and the Indian Ocean, to south central Africa, land of the Chokwe people. Living mainly in eastern Angola

FIGURE 36. Vanuatu sand drawing. (Photo courtesy of Vanuatu Cultural Centre)

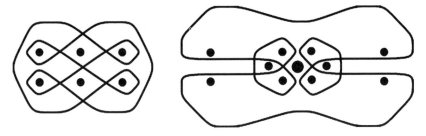

FIGURE 37. *Sona* sand drawings: the Antelope's Paw (left) and the Spider (right). (Images courtesy of Erik Demaine, MIT)

and northwestern Zambia, the Chokwe have a great artistic tradition in different media—pottery, wood, weaving, house decoration—and running through the art are designs and patterns that have their origins in sand drawing, or *sona* (Figure 37). Many of the designs are hauntingly evocative of those of Vanuatu, and the method of drawing, tracing a single continuous line in the sand to complete the pattern, is identical. The patterns help to tell (and remember) stories, fables, and riddles; they form games and record myths and history. The sand provides the medium in exactly the same way that it does in the desert of Australia and on the islands of Vanuatu.

In *sona,* the geometrical grid underlying the symmetry of the pattern is, unlike in Vanuatu, explicit. A matrix of dots, the size of which influences the final design, is made with a fingertip in the sand; then the finger traces out the pattern in a continuous, closed line, no part of which is ever traced over. The designs range from the simple to the extraordinarily complex. Like the old intelligence-testing puzzle where the challenge is to join up a three-by-three matrix of dots with a single continuous but nonrepeated line, the completion of a *sona* design requires drawing "outside the box"; but it also follows a number of fundamental mathematical rules and principles. At key points, the lines appear to reflect off invisible mirrors positioned both outside and within the matrix, and the positions and numbers of these mirrors determine whether or not a given matrix allows the completion of the pattern with a single line. The mathematics of this, together with the question, rather like the "traveling salesman" problem, of what is the shortest line that can com-

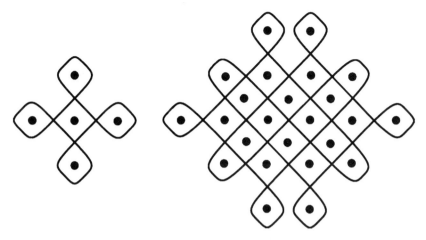

FIGURE 38. *Kolam:* the Anklet of Krishna (left) and the Ring (right). (Images courtesy of Erik Demaine, MIT)

plete the enclosure of the dots, is of interest to today's graph theorists and students of topology and symmetry—as well as "ethnomathematicians." The Chokwe artists and storytellers are undoubtedly unaware of the underlying algorithms—their designs in the sand speak for themselves and for the Chokwe.

Now, look at Figure 38—other *sona* designs, surely? But these patterns come from Tamil Nadu and are typical of that region and other parts of India, each of which has its own tradition and designs. These are *kolam,* patterns typically drawn outside the door of a house to welcome guests, facilitate prayers, assure the gods of the cleanliness of the house, and generally provide decoration. The designs are drawn by women, using rice flour, powdered rock, and colored sands. If the pattern is decoratively enhanced with colored images, it is called *rangoli.* During religious festivals, huge and complex designs are made in temples, often with the hope that wedding prayers will be answered. *Kolam* are commonly made by unmarried girls who keep this in mind, and who have learned the art from their mothers. Competitions are held for skill and new designs, and magazines are devoted to publishing new motifs. The underlying mathematics of matrices of dots and closed continuous curves is the same as that for *sona*—and the patterns reflect this.

Many of the *rangoli* designs are unique to the region and distinctly different from those of the Chokwe, but many are virtually identical. The continuity of design from central Australia through southern Africa to India must reveal something of our collective imagination and natural instinct for patterns, and for expressing ourselves using the Earth, especially sand, as our medium.

CIRCLES

Perhaps the most exquisite examples of artistic and spiritual expression through designs in the sand are the intricate mandalas of Tibetan Tantric Buddhism. Taking days or weeks to make, the mandalas depict harmony, a view of the universe, and a cosmic map; they provide healing, guides to meditation, and a reconsecration of the Earth and its people. And, like *milpatjunanyi, sona,* and *rangoli,* mandalas tell stories. The word *mandala* is Sanskrit in origin, roughly meaning "circle." The sand used to create one is made today from ground white stone, dyed in a rainbow of different colors. While traditionally the making of a mandala was a private ceremony enacted within the monastery, today Tibetan monks travel the world demonstrating their art. Once the preparatory rituals and prayers are completed, they draw, from memory, the outline of the design and then begin the painstaking process of filling it in with millions of grains of colored sand. They start at the center, slowly working outward, using sand poured meticulously from a small, narrow metal funnel, a *chakpur.* The *chakpur* has a rough surface, and the monk controls the flow of sand by causing it to vibrate by running a metal rod over the surface. Exploiting the natural behavior of the granular material, this controlled vibration causes the sand to flow from the funnel like a liquid (Plate 14 shows the construction of a mandala of Vajrabhairava, a wrathful nine-headed deity). The mandala can be 2 meters (6 ft) or more across and contain hundreds of exquisite and colorful images of traditional spiritual symbols—wheels within wheels. The entire design seems fractal, with each grain of sand ultimately a mandala in itself.

Indeed, this intrinsic reference in a mandala to scale and individual grains has been the inspiration for a collaboration between Victoria Vesna, an artist, and James Gimzewski, a nanoscience researcher at the University of California at Los Angeles. In 2004, in their installation *Nanomandala,* they projected onto a 2.5-meter

(8 ft) disc of sand a video that showed sand at gradually increasing and then decreasing scales, from the molecular structure of an individual sand grain to a completed mandala and back again. The installation was accompanied by sounds of the *chakpur* and of monks chanting, recorded during the creation of the mandala seen in the video. Visitors could participate by moving their hands in the sand as the projection continued, in part on the sand and in part on their own bodies. Vesna commented: "Inspired by watching the nanoscientist at work, purposefully arranging atoms just as the monk laboriously creates sand images grain by grain, this work brings together the Eastern and Western minds through their shared process centered on patience. Both cultures use these bottom-up building practices to create a complex picture of the world from extremely different perspectives."

A mandala illustrates the Buddhist view of the transitory and impermanent nature of things, and the sand, the medium for the design but itself a transitory and constantly changing material, provides the drama of the final gesture. When the mandala and the ceremonies are complete, the sand is swept up, the design instantly destroyed. The sand is placed in an urn, and half is presented to the audience; the other half is carried to a river and cast into the current, carrying the mandala's healing to the ocean.

The circular design, the image of circles within circles, labyrinths, wheels within wheels, is common across religions, cultures, and history. In North America, the Navajo use sand paintings in much the same way as the monks of Tibet, to provide healing and blessings, to restore order and harmony when nature seems out of balance. Like the Tibetans, the Navajo have hundreds of distinct paintings and designs, but also, in the same way, they are largely prescribed—the *creation* of the image is key, not individuality or innovation. Again like a mandala, a Navajo sand painting is accompanied by ritual and is a communal creation, the work being carried out by several artists, starting from the center and working outward, but following the Sun from the east, through south and west, to finish in the north (Figure 39). The dry painting is done on a sand floor, using colored sand together with other materials—stones, charcoal, gypsum, and flower petals. The patterns are symmetrical and contain images of the gods and human heroes, with the eastern side of the circle left open to allow these spiritual beings to enter the design.

After the painting has served its spiritual purpose, like a mandala it is destroyed.

FIGURE 39. Navajo sand painting. (Photo courtesy of Denver Public Library, Western History Collection; Mullarky Photo X-33166)

The sand is swept away in a sequence that is the reverse of the painting's creation, buried or cast to the four winds. The permanent sand paintings that are made on glue-covered boards add variations to the original designs, change the colors, or reverse the directions—the metaphor of impermanence of a true sand painting is at its heart, and to make it permanent would be a sacrilege.

AN IMPERMANENT CANVAS

Whether in the building of a sand sculpture, a mandala, or simply a beach doodle, the transient nature of sand as a medium has deep appeal. The sand castle is washed away, the doodle casually brushed clean, the mandala swept up. In New London, Connecticut, George K. Clarke creates "manhole mandalas" by filling the designs in the cast iron covers in the street with colored sand, to be washed away by the next rain. There are stories, some perhaps true, many no doubt apocryphal, of Pablo Picasso sketching in the sand on a beach; in one, he is pursued by a woman asking for a small sketch, and he obliges by drawing a picture in the sand. Transience and renewal on a grand scale underpin the allure of what has come to be called *land art,* earthworks that intentionally remind us of a constantly changing landscape.

Jim Denevan walks out onto a freshly washed Northern California beach, bends down, and draws a circle, the size of a coin, in the sand with his finger. More circles create a spiral nest, the outer ones growing larger. Using a driftwood stick as his paintbrush, he draws bigger circles in the sand, each one nestling with the previous; the design grows fractally. Denevan does all this freehand—there is no outline, no preliminary design; the artwork simply flows from his mind through the choreography of his movement. His work evokes a Japanese *karesansui,* popularly known as a Zen garden, with its contemplative design of raked sand. Denevan's design is monumental, ultimately occupying the entire width of the beach. He uses a large rake to highlight the outline and to fill in the spaces with texture (Figure 40). A few hours later, the tide destroys the art and renews the canvas. For Denevan, the transience is part of the art; it recognizes "some kind of truth about life—what is grand, or what is fragile. . . . Everything is transitioning into something else." Denevan's designs are diverse, ranging from huge perfect spirals, to representational images, to complex circles and linear shapes, suggesting a more fragile version of the Nazca lines, the gigantic figures in the Peruvian desert—themselves created by the removal of desert stones to expose the light-colored sand beneath.

The purpose and meaning of the Nazca lines are still a mystery, but if we view them as the oldest form of land art, then it is indeed a long tradition. As a modern form, land art began to blossom only in the 1960s with artists such as Michael Heizer, who sculpted circular designs with his motorcycle in the sand of the Nevada desert and excavated the giant trenches of his best-known work, *Double Negative* (1969–70). Heizer's father had been an archaeologist who took his son with him on his expeditions, including to Peru. Was he inspired by the earliest land artists of the Nazca? Another of the founders of modern land art was Robert Smithson, who is perhaps best known for his *Spiral Jetty* (1970), a gigantic work in earth and basalt; set in the Great Salt Lake in Utah, the spiral is sometimes visible and sometimes not, subject to fluctuations in the water level of the lake It was Smithson (tragically killed in an air crash at the age of thirty-five while surveying a site in

FIGURE 40. Examples of Jim Denevan's work. (Photos courtesy of the artist)

Texas) and his colleagues who demonstrated that art was something that could use materials from the Earth and exist outside an art gallery—producing works of art whose life, whose duration, could be subject to the rate of geological change, slow or rapid. Smithson introduced the term *earthworks* to describe land art, perhaps inspired by the books of science-fiction writer Brian Aldiss, whose 1965 novel *Earthworks* described a dystopian future of destroyed soil—robot tankers shipping sand from Africa to other parts of the world to replace soil ruined by human depradation.

Smithson often employed spirals in his work. His *Spiral Hill,* created in Holland in 1971, used white sand to highlight a ramp winding around a mound of black earth. What is it about circles and spirals that connects not just Denevan and Smithson but, apparently, all of us? Smithson also played with sand on a large scale indoors. His *Mirrors and Shelly Sand* (1970) is an 8.5-meter (28 ft) long mound of sand cut at regular intervals by fifty mirrors. By reflecting the sand, the mirrors extend and multiply it, making a mockery of scale.

Land art is very much about scale, from the initial displacement of a grain of sand, a pebble, a boulder to the creation of something monumental—or small. One of today's best-known land artists is Andy Goldsworthy, every one of whose works is profoundly anchored in the natural world. In *Fine Dry Sand* (1989), Goldsworthy used sand to describe itself, the sinuous undulating ridges and furrows emulating ripples, waves, and movement. Goldsworthy's *Dark Dry Sand Drawing* (1987), a design on a beach, prefigured Denevan's work. Other artists have used sand pouring from a backpack to draw meandering lines or have created human designs overlapping natural ones in the desert sands of the Kalahari. Enthusiastic and talented amateurs on the beach are also land artists, shaping repetitive designs like crop circles or competing to create images of celebrities. Where land art stops and playfulness begins is a matter of opinion, but all sand works are united by being more or less transient. They are preserved as photographs in books and on the web, but the photographs are not the art. The art, if it still exists, is defined by its context, its connection with the place and the materials, and its expression of continuous change.

Jay Critchley is another artist who has long been inspired by sand. His first works were temporary installations of "sand cars," which included a sand-encrusted station wagon and *Sand Family,* a car filled with sand. Visitors to Cape Cod can see—for now at least, but for how much longer is uncertain—his *Beige Motel,* an iconic

American 1950s one-story motor court (originally the Pilgrim Spring Motel) that Critchley has encrusted entirely with sand; the interior houses smaller installation works. The beige theme is evocative of what scientists, combining the spectrum of hundreds of thousands of galaxies, have determined to be the color of the universe. *The Beige Motel* typifies the symbolism of an impermanent canvas—the structure is due to be demolished. As Critchley says, it is "about temporarily capturing a very fluid substance for a moment in time, a motel about to be demolished, reduced to its earthly substances, mixing again with the sand."

PERMANENT CANVASES: PAINTING WITH SAND

Land art, as any land artist will tell you, is a messy business; sand is a material that has a habit of behaving as its own rules decree—and of getting anywhere and everywhere. This can also cause problems for the more conventional artist, particularly one whose enthusiasm for painting landscapes in the wild results in exposure to the penetrating whims of blowing sand grains. Georges Seurat at Gravelines, Claude Monet on the beach at Trouville, and Jan Vermeer on the rooftops of Delft involuntarily had sand grains mix with their paints. Also susceptible to granular intrusions was Vincent van Gogh. Painting *en plein air,* he captured the swirling landscapes and brilliant colors of Provence—along with sand grains borne into his pigment by the Mistral. Even in his early days in his native Holland, Vincent suffered from unwanted additions to his work. In a letter to his brother Theo in 1882, he wrote:

> All during the week we have had a great deal of wind, storm and rain, and I went to Scheveningen several times to see it.
>
> I brought two small marines home from there.
>
> One of them is slightly sprinkled with sand—but the second, made during a real storm, during which the sea came quite close to the dunes, was so covered with a thick layer of sand that I was obliged to scrape it off twice. The wind blew so hard that I could scarcely stay on my feet, and could hardly see for the sand that was flying around. However, I tried to get it fixed by going to a little inn behind the dunes, and there scraped it off and immediately painted it in again, returning to the beach now and then for a fresh impression.

Ironically—or perhaps as a reference—when Francis Bacon worked on his *Study for "Portrait of Van Gogh" III* (1957), some seventy years later, he intentionally added sand to increase texture in the already thick paint strokes.

Bacon was following in an illustrious tradition by intentionally using sand to add texture to his painting. Jean-François Millet, Georges Braque, Salvador Dalí, and Picasso all did so at one time or another. One of the great Surrealists, André Masson, used sand liberally in his paintings. Born in 1896, he was badly wounded in World War I, an experience that is cited as the origin of the conflict portrayed in much of his imagery. Like Joan Miró, Masson made "automatic" drawings, allowing a line to develop freely as, hopefully, a direct connection with the mind of the artist. Pursuing this goal in his painting, Masson would pour and dribble glue or gesso freely onto the canvas and then coat the glue shapes with sand, brush away the excess, and use these forms ("almost always irrational ones," according to the artist) to guide the rest of the work. His *Battle of Fishes* (1926), now in the Museum of Modern Art in New York, is a classic example of this technique. Masson emigrated to the United States in 1939, and his free-flowing style, with gesso and paint dribbled directly from the tube, is said to have influenced Jackson Pollock (who likened his own methods to those of Navajo sand painters).

In art museum shops today, you can buy a sand art kit for kids. The kit consists of paper, glue, multicolored sands, and a sand dispenser, together with detailed instructions, magic sticky sheets, and "special" sand art paper—the spirit of Masson lives on.

Kids visiting Alum Bay on the Isle of Wight, off the south coast of England, will drag their parents to a variety of carnival amusements that have grown up above the cliffs; among the less gaudy of these attractions is the Sand Shop. Vertical layers of sandstones in the cliffs below have been colored by diagenesis and the infiltration of mineral-rich waters, producing a variety of yellow, red, greenish, and ochre sands. In the Sand Shop are trays of these sands, which the children can pour into a variety of glass containers—lighthouses are popular—to make patterns, a tradition that began when Queen Victoria was presented with patterned sand in a bottle. The designs and execution of the Sand Shop pieces, even in the commercial works on sale, are crude, but not so the bottled sands of Andrew Clemens. Born in the mid-nineteenth century in Iowa and deaf from early childhood, Clemens developed an extraordinary skill. He knew of the place where the St. Peter Sandstone, originally deposited in a shallow sea 450 million years ago, had been stained into a rainbow of colors, from deep red through green to blue. Clemens collected the sands, cleaned and sorted the grains into different sizes, and, with a small hickory stick and a fishhook, placed them, one by one, into bottles to create—without adhesive of any kind—incredibly detailed pictures with shaded tones, geometric designs, and writing (Plate 15). Nothing like Clemens's work had been created before—or has been since. Used as gifts and custom greeting cards, sold for a few dollars, the surviving examples are now valued in the tens of thousands of dollars.

Sand thus provides the medium for art on all scales. One of the most extraordinary of today's sand artists who work on the small scale is Willard Wigan, who painstakingly carves from individual sand grains or fragments thereof. An elephant, for example, occupies a fraction of the area on the head of a pin. One hazard of this art is that it can easily be lost through inhalation or sneezing, a serious loss when months of work are involved and the value is significant: his collection recently sold for over twenty million dollars.

SAND ART IN THE DIGITAL AGE

In 1949, a young graduate student in engineering in Miami, Joe Woodland, was determined to develop a method for automating the checkout process at supermarkets. Down at the beach one day, Woodland was idly tracing patterns in the sand, based on the only code he knew, Morse code. His fingers elongated some of the dashes, and the bar code was born. Woodland was ahead of his time: everyday use of the bar code would emerge only much later. Nevertheless, one of the icons of the digital age owes its origin to sand as a medium.

Today, the ability of sand to serve as a three-dimensional, malleable canvas for capturing and displaying information is thoroughly integrated into the digital age. As we have seen, sand has lent itself to this use for millennia—*milpatjunanyi* and *sona* are living traditions, and sand tables were not only tools for teaching writing but, on a large scale, continue to be used for military planning. In one form of contemporary sand art, images are continuously sculpted and reformed on a glass plate by the artist's hands, the whole process illuminated from beneath and magnified and projected. The art is intended to be *watched*—there are countless video files of these works on the internet. But the digital relationship runs deeper. The installation *Nanomandala* discussed earlier is but one example of using a surface of sand as a medium for digital imagery.

Mariano Sardon is an Argentinian installation artist who has also used sand and digital imagery in his work. Now a professor at the Universidad Nacional de Tres de Febrero, he began as a plasma physicist at the University of Buenos Aires. Starting in 2003, Sardon developed an installation called *Books of Sand*, after "The Book of Sand," by fellow Argentinian Jorge Luis Borges, the story of an infinite, constantly changing, and deeply disturbing book. Sardon's installation is interactive: the viewer's hands manipulate sand in large glass cubic containers, and at the same time the movements of hands and sand are processed to instantly feed back ever-changing text fragments of Borges's story plucked from web pages; the text fragments are projected on the sand and the viewer, fluidly changing, appearing and disappearing. In Sardon's words: "The cubes enclose in a confined space a fragment of the infinite information flowing through the web. It is immeasurable and never ending, like the particles that make up the sand and the codes that form the text. In 'Books of Sand' the viewer has the possibility of handling what is immaterial, can grasp an instant of it in the fist of a hand." The effect is compelling—ethereal, kinetic, shape-shifting, luminous. The viewer, sculpting, pouring, and scattering the sand, appears to *create* the text.

The journey from science to artistic expression is a common one in the digital age. For some time, MIT has run a Media Lab, where recently *SandScape,* an installation evocative of Sardon's work, has been developed: as three-dimensional digital data (for example, topographical information) are projected onto a sand table, the viewer sculpts the sand in response to the data, which, in turn, respond to their manipulation. Sand, as for the native people of Australia, Vanuatu, and Angola, continues its role as a tangible "tactile interface."

Among other things, Bruce Shapiro builds "shimmibots" and "geyserbots," and creates works with titles such as *Ribbon Dancer, Stratograph,* and *Sisyphus.* Shapiro began his career as a medical doctor but has spent the past fifteen years exploring the relationships between motion control—which he defines as "techniques for or-chestrating the movement of machinery and objects"—and art. Salvaging parts from industrial robotics and automation, he builds machines that do extraordinary things, and many of them do these things with sand. His *Stratographs* are containers that very slowly fill up with colored sand, creating images over the course of an exhibi-tion, perhaps several months in length. His "shimmibots" take Chladni patterns (chapter 2) several steps further, using simple and complex vibrations to create liv-ing and evolving patterns in sand. The various versions of *Sisyphus,* of which there are several generations, take the form of large (up to 3 meters, or 10 feet, in diam-eter) circular sand tables; in the sand lies a small steel ball, set in motion by a pro-grammed device that drives a magnet beneath the table. The motion of the mag-net is controlled by the input either of an original design or of algorithms, equations that *Sisyphus* translates into graphic reality in a way similar to the gen-eration of fractal images from mathematical equations. The size of the sand grains is critical: Shapiro uses very fine sand to optimize the movement of the ball (whose task, as a result, is less onerous that that of the original Sisyphus, who endlessly pushed a boulder uphill) and the preservation of the terrain it creates. Figure 41 il-lustrates the results of *Sisyphus's* labors; among its stunning diversity of designs are nested circles and spirals—shapes of a seemingly universal, basic appeal.

This is, again, a dynamic work to be *watched,* not simply looked at—the ball moves sedately but determinedly, as if moved by an invisible but methodical hand. It can take hours or days to complete a sand "etching," after which, in the spiritual tradition of sand as a transient medium, *Sisyphus* destroys its own work.

One of the latest incarnations of *Sisyphus* is on permanent display at Techno-rama, the Swiss Science Center, near Zurich. It is one of an extraordinary and com-pelling variety of exhibits at the museum that translate, through technology, ma-terials into art—a fascinating number of which exploit the peculiar and fluid behaviors of granular materials that were described in chapter 2. Technorama, which refers to *Sisyphus* as "a kind of icon" of the exhibition, carries a series of videos on the web—appropriate for art that is to be watched.

Given the contemplative state it invites in its viewers, *Sisyphus* has understand-ably been described as evocative of the raked gravel patterns in Japanese Zen gar-

FIGURE 41. *Sisyphus* at work. (Photo courtesy of Bruce Shapiro)

dens. The performance artist Mona Hatoum has exhibited a work with similar associations: in *+ and -* (1994–2004), a large circular sand pit is crossed by a metal rake that is serrated on one half, smooth on the other; the rake slowly rotates, simultaneously creating and destroying a set of grooves in the sand. It is an automated Zen garden and shares spiritual kinship with *Sisyphus*.

Dynamic installation works all over the world explore the interface between art and science by harnessing the strange behaviors of sand in air and water. Every image, every form they create, is unique—and often temporary. The allusive properties of sand are explored in projects like the computer-driven sand etchings of Jean-Pierre Hébert (Shapiro's early collaborator), the interactive installations of Ned Kahn *(Aeolian Landscape, Fluvial Storm)*, and Martha Winter's fusing of geology and geometry into three-dimensional canvases of pigmented sand. When the nineteenth-century French mathematician Jules-Antoine Lissajous set up his cord-

and-funnel apparatus to trace out in poured sand the ellipses that now bear his name, he little guessed how much more his medium could express.

READING THE SAND: DIVINATION

If certain Taoist elders or ancient Arabic visionaries or Dogon priests from Mali were to watch *Sisyphus, SandScape, Books of Sand,* or Jim Denevan creating a sand work, they would know exactly what was going on. They would see not art, however, nor technology, but the future.

Geomancy, the art of divination, is a word that derives from the translation of the Arabic term *ilm al-raml,* the science of the sand. In a wide variety of cultures, sand has been a medium for telling the future. A shallow bowl of sand can be used, the surface smoothed; the person whose fortune is to be told makes shapes and patterns in the sand, the expert interpretation of which defines his or her future. Alternatively, the edge of the bowl can be tapped with a stick and the resulting patterns— vibrated granular materials again—can be interpreted.

In original Arabic geomancy, the diviner used and developed dots drawn in the sand on a board—or simply in the desert—by the person seeking advice. A twelfth-century automated interpreter, a beautifully ornate brass mechanical "calculator" from Damascus, was created to facilitate interpretation of the patterns. The Arabs brought the art of geomancy to Africa, and the tradition continues. Today, complex patterns of lines and dots in the sand—often, like *sona* and *rangoli,* displaying fundamental mathematical qualities—are used to answer questions and divine the future. In the Dogon culture of Mali, the priest sets out an elaborate pattern of drawings, lines, and piles of sand, and invokes the sacred fox to visit during the night. If the fox obliges, the pattern of its tracks around the priest's arrangement is interpreted.

In certain forms of Taoism, divination is accomplished by using sand as the medium in which spirits and deities write. Fine white sand in a large sand tray is smoothed and preparatory rituals are completed. The expert medium enters into a trance, takes a stick, and begins the incantation to seek assistance. If the incantation is successful, the thoughts of the spirit are transferred through the medium and the stick into sand writing. The words and symbols are continually recorded and the sand resmoothed so as not to interrupt the procedure.

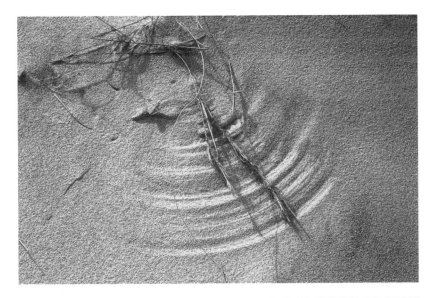

FIGURE 42. Sand designs of wind and water. (Photos by author)

FIGURE 43. The Earthquake Rose. (Photo courtesy of Norman MacLeod)

THE EARTH RESPONDS

Sand is an infinitely expressive medium—and muse—and is widely employed across geography, cultures, and time. It can be used to tell stories of the past and of the future, and to portray the constantly changing present. And as we have seen in the rivers, beaches, and deserts of the world, sand is also the medium for the expression of the patterns, microscopic and gigantic, of the Earth itself, of nature, whose transient designs are themselves art (Figure 42).

On at least one occasion, nature has created its own Zen garden, emulating *Sisyphus.* In February 2001, in a shop in Port Townsend, Washington, a desktop toy— a sand-tracing pendulum—was set in motion. It was tracing out a simple design in the sand when a strong earthquake began shaking the building. When all had calmed down and people returned, the Earth had traced the record of its shaking in the center of the sand, sometimes now referred to as the Earthquake Rose (Figure 43).

9

Servant

Sand in Our Lives

A foolish man, which built his house upon the sand.

Matthew 7:26

BUILT ON SAND

The fate of Ozymandias, the much-debated significance of Lucky's sand-filled suit-case in *Waiting for Godot,* pockets full of sand, lessons made of sand—gloom, despair, and pointlessness are the messages. Sand has come to be used as a symbol of the fragile side of the human condition: foolishness, futility, worthlessness, heart-ache. But as we have seen in the previous chapter, sand can also be a source of in-spiration and creativity. It has a complex role. In reality, as Jorge Luis Borges wrote: "Nothing is built on stone; all is built on sand, but we must build as if the sand were stone" (*In Praise of Darkness,* 1974). And so we do. If a wicked fairy were to wave her magic wand and remove everything around us that owes its origin to sand, things would look disturbingly different. If the wand waving occurred at night, the effects would be less easily seen because the lights would have gone out.

We are a society built on sand. Given that sand is one of the most ubiquitous materials on the planet, this should not be surprising; but the sheer diversity of the ways in which sand plays a role in our lives (beyond the purely recreational) may be unexpected, as sand often acts behind the scenes. The diversity of sand's roles makes this a difficult chapter to construct—to spin a coherent story from such a huge cast of disparate characters would be strained, even if it were possible. I shall therefore fall back on that old, trusted structure for compendia—the A to Z. Over the following pages is a miscellany, a selection of topics that highlights the leading roles of sand and alludes to others: the obvious and the invisible, the macroscopic and the microscopic, the serious and the whimsical.

A

Sand plays a leading, if often primitively simple, role in the class of industrial commodities referred to as *aggregates.* These are all the rock particles and lumps—whether sand, gravel, crushed rock, or recycled materials—that are used to build and weight down things. Roads and concrete are the primary beneficiaries (see *concrete*). Traditionally, the United States has been the largest producer of aggregates in the world, churning out well over a billion tons of sand and gravel for the construction industry every year. However, given the fevered rate of construction in China, including its record consumption of concrete, it is quite likely that the global hotbed of aggregate production has shifted: production levels in any country are intimately linked with its economic health and the exuberance of its construction industry. Industrial sand production relies largely on the work of rivers and ice sheets eroding, transporting, sorting, and dumping deposits of an attractive commercial quantity and quality of sand. Extraction is done almost entirely by surface mining, whether on land or the seabed: marine deposits are an increasingly important source, removed by dredging or suction (see *islands* and *Korea*).

As anyone who has experienced a sandstorm or observed objects subjected to blowing sand can testify, the material is highly *abrasive.* Sandpaper represents the familiar exploitation of this property around the home. Sandpaper has been used at least since sand and crushed shells were glued to parchment in China seven hundred years ago; mass production began with glass paper in England, and sandpaper was patented in the United States in 1834. Today, much of the abrasive material used in sandpaper is not quartz but aluminum oxide and silicon carbide, but it's all sand, by definition, because of its size (and, anyway, silicon carbide is manufactured from quartz sand). Sandpaper grit sizes correspond closely to Wentworth and Udden's scales for natural sands down to the "microgrit" category, in which the abrasive material is finer than very fine sand.

The ancient Egyptians used sand in combination with bronze saws to cut stone for their monuments—sand grains have been found embedded in abandoned cuts. The method was described by Pliny, and it was employed for centuries. Samuel

Pepys, in an entry from his famous seventeenth-century diary, related this example of its use:

> I staid a great while talking with a man in the garden that was sawing a piece of marble, and did give him 6*d* to drink. He told me much of the nature and labour of the worke, how he could not saw above 4 inches of the stone in a day . . . and after it is sawed, then it is rubbed with coarse and then with finer and finer sand tille they come to putty, and so polish it as smooth as glass. Their saws have no teeth, but it is the sand only which the saw rubs up and down that do the thing.

Today's grit-impregnated saw blades are simply the latest examples of the technology. Sandblasting replicates a natural sandstorm, but in a controlled way, using the same materials as sandpaper. I spent a good part of my youth in Manchester, England, where the architectural pride of the city is the gigantically gothic Victorian town hall. As a landmark of the Industrial Revolution, it was always thought by the citizens to be naturally black—until it was cleaned, sandblasted to reveal the glowing 300-million-year-old sandstone with which it was faced. Countless buildings around the world have been refreshed, if somewhat brutally, by sandblasting. Frank Gehry's dramatic Disney Concert Hall in Los Angeles was recently ignominiously subjected to the treatment to dull its undulating stainless-steel structure, which was too blindingly reflective.

Sandblasting has replaced hand carving as a way to inscribe gravestones and is a standard method of precision glass etching. It can also be a controlled method of sculpting stone in preparation for the final, detailed work of masons. The 1920s Barclay-Vesey Building in New York, the first art deco skyscraper in the city and a classic piece of architecture, was severely damaged on September 11, 2001, and the replacement panels for its sculpted exterior are works of art by expert sandblasters and masons.

There are also some bizarre applications of sandblasting. If you want your skin to have that glowing, refreshed look, you can try "microdermabrasion," where a jet of very fine sand is used to remove dead skin cells and unblock pores. But, tragically, the production of sandblasted jeans causes workers in unregulated factories around the world to suffer from the malign and deadly side of sand—silicosis. This debilitating and incurable lung disease, first recognized in stonecutters in the eighteenth century (but also seen in ancient Egyptians through autopsies of mummies), is caused by the inhalation of fine silica sand and dust.

B

Before the invention of blotting paper (first mentioned in English in a fifteenth-century text), *blotting* sand was used to absorb ink on manuscripts; even today, every desk in the United States Senate is equipped with an inkwell and a sand shaker for blotting. Grains of sand not blown off by the writer occasionally fall out of old manuscripts. (Incidentally, quill pens were "cured" by immersing them in hot sand.) The blotting ability of sand, its capacity to soak up liquids, is a reflection of the physics of surface tension on the grains. It was put to great use in the gladiatorial arenas of ancient Rome to soak up blood and other fluids. This characteristic of sand is used today to handle emergency spillages in laboratories and factories. Coats of blotting sand are now required on new road asphalt (which itself contains sand).

Cows whose bedding is sand have far lower levels of bacteria than those bedded in sawdust; the sand is almost equally effective as an absorbent but more sterile than sawdust. There are machines specifically designed to process the sand and remove the manure.

Sand not only absorbs liquids but also absorbs energy. Sand-filled barrels were developed for use around motor-racing circuits after a particularly bad accident at Le Mans in 1955. The energy-absorbing character of sand provides one of the more frustrating elements to the game of golf, the numerous different ways in which a golf ball can bury itself in the granular material of a sand trap stimulating a range of golfing terminology—and expletives. Sand is what long jumpers jump into. And, on a more subtle level, aficionados of the production of fine sound use the damping qualities of sand to stabilize speakers and other audio equipment.

C

When the wicked fairy waves her magic wand, a large part of the fabric of modern cities and towns will simply disappear—your vanished diamonds will be the least of your concerns. The *construction* business is entirely dependent on sand, and our modern cities are monuments to it. The use of sandstone as a building material is obvious. In Washington DC, the White House is white because it was painted to cover the poor-quality sandstone that the Scottish masons had to work with (the Scots being used to Edinburgh, a city built of much older, more solid rock). The original Smithsonian Institution building is made of sterner stuff—Triassic red sandstone. The Pentagon, however, is made of *concrete,* for which nearly 700,000 tons of sand were dredged from the Potomac River.

The recipe for basic concrete is simple and has been around for a long time. The ancient Egyptians knew how to make it (there is a lively debate as to whether the pyramids, at least in part, are made of concrete), and the Romans perfected the formula. The fundamental ingredients are around 75 percent sand and gravel, 15 percent water, and 10 percent cement. The cement, cooked from materials such as limestone and clay, is the chemical glue; the hardening of concrete is not simply due to drying but involves complex chemical reactions. The physical characteristics of the sand, its size and shape, influence the properties of the concrete, but because of the importance of chemistry, the composition of the sand and the other ingredients is critical. The wrong impurities will ruin the quality of the concrete (which is why the trade in salt-encrusted sand by villagers in *The Woman in the Dunes* is illegal).

The global demand for concrete is massive: after water, concrete is the most consumed material on Earth. Every year, the equivalent of more than 400 million dump trucks of concrete is transported to construction sites. Every man, woman, and child on the planet "consumes" around forty times their own weight in concrete per year. Which is, of course, an average—for residents of the Western world, it's much more, despite the fact that around half the world's concrete production and consumption today is accounted for by China.

Entire cities have grown into icons in concrete. Where did so many modern buildings in New York City come from? Glaciers and Long Island. In 1865, mining began on the northern shore of Long Island to collect sand washed out from retreating ice age glaciers. Immigrant workers from Europe, many from Sardinia, first hauled sand with wheelbarrows; the excavations grew with mechanization, and eventually the cliffs and the landscape were leveled. Port Washington was the center of the business, as endless convoys of barges carried the sand to Manhattan. The last sandpit closed in the 1990s, by which time more than 200 million tons of sand had been excavated to build the city—bridges, highways, the Empire State Building, the

Chrysler Building, and the World Trade Center. As Al Marino, a worker in the pits, is quoted as saying: "The sand we got in the Port Washington sandbanks is fantastic. . . . It has life in it. It's the best sand you could get for making concrete—just the right combination of coarse and fine grains. You go to the beach and you take that sand. And if you make concrete out of that it would fall apart, because there's no life in it" (Elly Shodell, Port Washington Public Library). The construction of Kennedy Airport took over two million dump trucks of sand from Jamaica Bay at the other end of Long Island—a big hole in the sediment budget.

Concrete is used in the construction of prosaic and functional buildings—and in fine architecture, from Antoni Gaudi's Sagrada Familia cathedral in Barcelona (sometimes likened to a sand castle) to Frank Lloyd Wright's Solomon R. Guggenheim Museum in New York. Even Gehry's soaring sculptures in metal and glass need concrete to frame and underpin them.

Common and basic though it is, concrete lends itself to technological innovation. Traditional additives have made it faster curing, lighter weight, stronger, and resistant to corrosion, and there are some extraordinary modern developments. Among Thomas Edison's many visions was one of cheap and simple concrete housing construction; his ideas were, again, ahead of their time, with public enthusiasm notably lacking, but his company did provide the concrete for New York's Yankee Stadium. Today, materials science has taken up the challenge, and if you soak old newspapers and magazines to make papier-mâché, then mix it with sand and cement, you have fibrous cement, a building material that is lightweight, cheap, strong, and capable of being sprayed or sawed. The material is highly fire resistant and, because of the structure of the sand grains and fibers, has excellent insulating and thermal properties.

Add even small amounts of steel or carbon fibers to concrete and it will conduct electricity—it becomes a material with the ability to monitor itself. A road made of conductive concrete can be warmed up, using electrical heating, to prevent the formation of ice, averting the use of sand or salt. Traditional concrete is brittle and cracks easily; modern concretes can be flexible, again through the addition of fibers, and made translucent through the addition of glass optical fibers. Self-cleaning concrete, harnessing the chemical activity of titanium dioxide (itself often derived from sand), has been developed in Italy, and it seems that it could actually clean up polluted air: titanium dioxide becomes chemically reactive when exposed to light, absorbing pollutants.

Specialty concretes also figure on the domestic front. Colored and polished "designer" concrete has become the material of choice for kitchen floors and countertops, benches and desks, the constituent sand grains exposed, if desired, to provide the feeling of a beach or a sandstone. The results are stunning, but the customizing process is often labor-intensive and expensive.

The main problem with all types of concrete lies in its production, particularly in the manufacture of cement: cement making may account for up to 10 percent of global carbon dioxide emissions. Alternatives to cement have proved difficult but not impossible to develop, and recycling, particularly of glass, plays an increasingly important role in concrete manufacture. On many modern construction sites, gigantic machines can be seen imitating the processes of nature, spewing out the ground-up material from the demolition of a previous building to create artificial aggregate.

D

The use of sand in building extends far beyond the basics of construction. Key materials for *decoration* also rely on sand and its derivatives. For centuries, many types of plaster have used sand to provide cohesion. I looked closely at the remarkably preserved plaster surface of the walls of a Roman temple in a Saharan oasis, and the light caught the coarse grains of embedded sand. A professional microscopic analysis of the sand grains in the plaster of ancient churches will yield stories of their age and construction. The delicate art of frescoes relies on the layering of plaster with increasingly finer sand content—the sand must be angular to maximize its binding effect.

Paint is a decorative material that seems far removed from sand, but this couldn't be further from the truth. Quartz sand, or silica (silicon dioxide), is the basic source of silicon, and silicon is a magic element, convertible both physically and chemically into a huge variety of useful substances. Ground to a microscopically fine powder, silica is used as a filler to control the physical properties of various products and to add bulk to them. Among these products are plastics, rubber, adhesives—and paint, the chemical inertness and hardness of silica making the paint more durable. Perhaps the simplest decorative application of silica is *smalt,* a coating long used to provide the background to painted lettering on signs; its base is either ground-up colored glass or simply sand, with pigment added.

Silicones—complex synthetic, silicon-based molecules—are the key ingredients in specialty paints. Silicones in general are remarkable materials and play a vast number of roles in our lives, in cookware, sealants, cosmetics (see *personal care* and *pharmaceuticals*), and anatomical augmentations, to name a few.

White paint, whether gleaming on your kitchen appliances or coating the White House, owes its whiteness to titanium dioxide, which was substituted some time ago for toxic white lead as a pigment. The titanium is extracted from the minerals rutile and ilmenite (oxides of titanium and titanium and iron), which are mined as sand grains concentrated naturally by ancient rivers and waves (see *jewelry*). More than half the titanium dioxide manufactured goes into paint; some of the rest goes into toothpaste, sunscreen (it blocks ultraviolet rays), paper, inks, food coloring, and, as we have seen, clever concrete.

D is also for *defense,* not only in a military sense—of hot sand poured onto the enemy from castle battlements and the construction of earthworks and trenches—but in the sense of defending against the forces of nature. Sandbags have been used as emergency barriers against floods for centuries. Massive sandbags were used to plug gaps in the levees caused by Hurricane Katrina—unfortunately, after the major flooding had already occurred. Improvements on the traditional technology of flood-proof barriers have been developed, but the majority of them are still filled with sand.

Of course, sand can not only keep out water but smother fire too. It also provides a defense against termites, which are unable to tunnel through loose coarse sand; under houses or foundations, around fence posts and telephone poles, sand forms an effective barrier.

E

Electronics. Where to start (and stop) on a subject that could fill a book in itself? Perhaps with a common oversimplification. "Sand to chips" is a popular conception and, indeed, in many ways it is as simple as that. The chips—the microprocessors in our computers, microwave ovens, mobile phones, and endless other electronic devices—are based on silicon, and silicon—a lot of it at least—comes from sand. But only from quartz sand, and only after undergoing an incredibly sophisticated and complex series of processes to make silicon of the purity required for electronics. You can, if you wish, make silicon at home, using a crude imitation of the real extraction process: heat a mixture of quartz-rich beach sand and magnesium powder to a considerable temperature; the magnesium tears the oxygen atoms away from the silicon and among the detritus in the bottom of your test tube will be some silicon—although it will be extremely impure. Electronics-grade silicon has to be at least 99.99999 percent pure—referred to in the trade as the "seven nines"—and often it's more nines than that. In general, we are talking of one lonely atom of something that is not silicon among billions of silicon companions.

Then there's the weird chemical behavior of silicon. Oxygen atoms have an immensely calming effect on it; silicon dioxide in the form of quartz, as we have seen, is one of the most durable substances on Earth. But silicon on its own, a brittle, gray, metallic material, is chemically promiscuous, reacting vigorously with almost anything. So why is silicon *the* material for making computer chips? The answer lies in another of its strange properties: it is a semiconductor. Most materials either conduct electricity or don't. (The latter are insulators, of which glass, primarily made of silica, is, ironically, an excellent example.) Silicon is one of the rare materials that is neither wholly a conductor nor wholly an insulator; it conducts electricity only under certain circumstances, which, importantly, can be controlled. Silicon is prepared to accept into its structure atoms of other substances that determine its electrical conductivity, and those foreign atoms can be introduced artificially (a process called *doping*); hence its role in electronics.

Clearly, when a substance of such extraordinary purity is required, the purer the starting material, the better—which is why any old sand won't do. So what are the sources of the raw materials? In thinking about this book, I had in mind the geological perspective—what are the origins of the sand grains that provided the means to enable the computer on which I am writing? A simple question, but surprisingly difficult to answer. First of all, the public is understandably more interested in the technology of making the silicon chip than in its provenance in a sandbank by a river somewhere. Second, the absolute volume of semiconductor-grade silicon is only a small fraction of all silicon production. But more important, a small number of companies around the world dominate the technology and the market, and while their literature and websites go into considerable and helpful detail on their products, the location and nature of the raw materials seem to be of "strategic value," and thus an industrial secret. I sought the help of the U.S. Geological Survey, which produces comprehensive annual reports on silica and silicon (as well as all other industrial minerals), noting that statistics pertaining to semiconductor-grade silicon metal were often excluded or "withheld to avoid disclosing company proprietary data." The survey staff were, as always, extremely helpful, but were themselves perplexed that such an apparently simple question was not simple to answer. They put me on the trail of a number of sources, but telephone and e-mail inquiries did not shed a great deal of light. What I could deduce is this: the common source of silica for manufacturing the high-purity grades of silicon used for, among other things, silicon chips is not loose, unconsolidated sand from a beach or river sandbank, but sand that has already been ultrapurified by nature: quartzite.

Quartzite is a rock that was originally a silica sandstone; it has been so deeply buried in the Earth's crust, cooked by such high temperatures and pressures, that many of the impurities have been distilled out and the sand grains completely annealed and welded together. Hit quartzite with a hammer and it rings like a bell because of its hardness, purity, and uniformity. Hit it hard enough and it breaks *across* the ghosts of the grains, not around them. Quartzite can be well over 99 percent pure silica. Grind it up to a powder of a consistent grain size and it's a good starting point for making silicon.

As we saw in chapter 7, the story of the Appalachians is divided into three dramatic episodes of mountain building, separated by periods of major erosion and sediment deposition. During the dramatic episodes, sands were buried and cooked into quartzites; during the interim periods, those quartzites were elevated to the surface and eroded. Rivers carried pebbles of quartzite down from the mountains, to be buried and once again lithified. Today, rivers carry quartzite pebbles, along with sand, gravel, and mud, out of the heart of the Appalachians, down toward the sea, as we saw in chapter 4. The Coosa River, for example, originates close to the border of Georgia and Tennessee and, together with its tributaries, drains a large and geologically diverse area of Appalachian rocks, including quartzites. It crosses from Georgia into Alabama, where it joins the Tallapoosa to become the Alabama River. As it does so, it enters the broad coastal plain, slows down, and dumps its sediment load. Large volumes of aggregates for all kinds of industrial purposes are extracted in this part of the country from the sediments of the Coosa and other rivers, but among those everyday aggregates are pebbles of great value— pure quartzites from the kitchens of the Appalachians. And those, I believe, are one source of raw material for high-grade silicon. So it's true that computer chips are made from sand—but sand that was first deposited several hundred million years ago.

Only slightly less pure than the "seven nines" material, so-called metallurgical-grade silicon has a host of uses, being an ingredient in specialty steels, alloys of aluminum, and silicones. To make this high-purity silicon, powdered quartzite is burnt with charcoal or wood at temperatures over 1,700°C (3,000°F); the oxygen atoms are seduced into eloping with the carbon (uniting to form carbon dioxide) in sequential steps of ever-increasing purification. A minute fraction of a batch of metallurgical-grade silicon is subjected to the ultimate purification. The specific technologies are highly proprietary, but the common method is to make the silicon into a liquid, convert the liquid to a gas, purify the gas, and condense it into the "seven nines" material—polysilicon. Further steps are then needed to fashion it into formats for solar cells or the brain of your microwave oven. Much of today's production of basic silicon metal is done in China.

Pure silicon is the prime ingredient for tens of thousands of substances with hundreds of thousands of uses, yet it still requires an energy-intensive and polluting technology that has been used for thousands of years: smelting. Is there no better way? Research into using nanotechnology or low-temperature chemical catalytic approaches demonstrates that there are, indeed, better ways, and ways that can even be based on everyday sand. And what of silicon itself—are there substitutes? While there is no shortage of sand or quartzite, there are competing demands for and a limited supply of high-purity silicon. And the silicon chip itself, while constantly being improved, is not perfect—it has limitations in terms of efficiency and energy loss. Two of the major corporate players have recently announced that the element hafnium performs better than silicon in this application, losing less power and allowing smaller and smaller chips. Hafnium? It's a somewhat obscure element of, so far, limited application—worldwide production is only around fifty tons per year, much of which goes into control rods in nuclear power plants and the rest into sophisticated super alloys for jet engines. And from where do we get hafnium? It occurs in partnership with its sister element, zirconium, in the mineral zircon, which is extracted from sand (see *jewelry*).

F

The manufacture of silicon uses the ancient technology of the *foundry:* smelting. Sand has historically played a key role in foundries, providing a material for casting the metal. Sand, combined with clay or chemical binders, is shaped around a pattern so as to contain the hollow form into which the molten metal is poured, and sand "cores" are used to form recesses and cavities in the final metal shape. After casting, the cores, along with the rest of the sand forming the cast, are simply removed. This is an application where round sand grains seem to be best, selected for the appropriate size. High-precision casting can be achieved this way, and while other technologies have been developed, this age-old method continues to be used today.

Among the many valuable qualities of sand are its ability to contain fluids, its porosity (see *reservoir*), and the ease with which fluids flow through it—its permeability. The unit of measurement that is used to describe the permeability of sand is the darcy, and the process of fluid flowing through sand operates according to Darcy's law. Henry Philibert Gaspard Darcy lived a short life, from 1803 to 1858, but he revolutionized the engineering of municipal water supplies. He was employed by the French Corps of Bridges and Roads and ultimately became its inspector general. He was personally responsible for designing and building a radical new water-supply system for the city of Dijon, the construction of which depended on his quantification of the physics of fluid flow. In the course of this, he analyzed and put to use the behavior of sand as an effective *filter,* studying how the spaces between the grains capture solid materials but allow the clean water to flow through. Water treatment plants (as well as septic tanks and swimming pools) all over the world still depend on sand as a filter. The filtering process often uses additional materials, such as carbon and other chemicals, but it's the sand that provides the basis for the job.

When lethal levels of natural arsenic were found in shallow domestic water wells not long ago in Nepal, Bangladesh, and other parts of Asia, the solution was sand filtration: "slow sand" filters, with iron oxide coating the grains, and "bio-sand" filters, cheap, simple and requiring no electricity, removed the arsenic. In Australia, iron-rich sand grains have been found to contain tiny pits as a result of natural weathering; the pits, just nanometers across (a million nanometers equal four hundredths of an inch) are exactly the right size to capture industrial chemical pollutants. Old principles, new technology (see *nanotechnology*).

G

Glass: the stone that flows. The technology of glassmaking is an ancient one. The earliest glassworks yet discovered were recently excavated in the Nile Delta; they date from around 1250 B.C., when Rameses the Great ruled the empire. But the belief that the technology originated in Egypt is wrong, as is, sadly, Pliny the Elder's satisfying fairy tale of serendipitous discovery. He describes, in his *Natural History,* how Phoenician traders with a cargo of natron—or soda (sodium carbonate)—perhaps from the desert lakes of the Sahara, had put in for the night on the eastern Mediterranean coast. Unable to find adequate rocks to support their cooking pots over the fire, they resorted to using some blocks of their cargo. Whatever their dinner recipe was, they had unwittingly assembled the ingredients for glass: the soda lowered the melting point of the beach sand, and out of the fire flowed streams of translucent liquid. While this story is undoubtedly apocryphal, the discovery of glass was probably a similarly serendipitous conspiracy of circumstances, but in Mesopotamia. Simple glass beads from around 2300 B.C. have been found in Iraq and Syria and in the Caucasus. The technology developed rapidly from there.

Glass beads were the earliest products of this technology, simple but nevertheless of an infinite variety of colors and designs, used for decorating, purchasing, and warding off evil. The age of exploration of the New World and the requirements for gifts and barter with the indigenous inhabitants stimulated a major industrial expansion. Christopher Columbus and Hernán Cortés carried large quantities of beads in their cargoes. For the Hudson's Bay Company, beads were the basic currency of the fur trade, and even though the story of beads being bartered for Manhattan is probably untrue, William Penn did use beads to seal the agreement on the land that would become Philadelphia. Tragically, currency in glass beads also drove the slave trade. Much of this industry was based in Venice, a glassmaking center now for over a millennium. By the 1800s, the city's manufacturers were exporting more than six million pounds of glass beads every year.

The development of glass is a tangled tale of chemistry, alchemy, invention, secrecy, serendipity, ideas, needs, breakthroughs, and the geology of the raw materials. To melt sand on its own requires temperatures in excess of 1,600°C (2,900°F), far beyond the capability of traditional wood-burning furnaces. The melting point has to be lowered to a practicable temperature, and this is where the natron plays its critical part in the story. Soda has a much lower melting point than sand and acts as a *flux,* an additive that makes silica sand meltable at realistically achievable temperatures. An alternative to natron, which is gathered from dried-out lakes, is the ash obtained from burning certain plants or seaweed, and some species from the eastern Mediterranean were found to be ideal. But care must be taken: too much flux and the resulting glass is unstable, and adding even the necessary quantity of flux makes the glass soluble in water, hardly a desirable characteristic. To stabilize the glass, calcium in the form of powdered limestone has to be added, and it is entirely possible that this was first discovered by accident through using silica sand that also contained fragments of seashells. The resulting concoction is soda-lime glass, the everyday kind of glass that has been used for the last three thousand years. To this brew, other components can be usefully added, each conferring its own special characteristics.

Though colorless and completely transparent glass is today the norm, colored glass is often desirable, as long as the color can be managed and predicted. Adding gold in very small concentrations creates a deep ruby color; cobalt produces blue; copper oxide, turquoise; and pure copper, a dark red, opaque glass. Add a myriad of other minor, special ingredients, and the physical character of the glass, its optical, thermal, and electrical properties and its strength, are profoundly changed. It is this chemical sleight of hand that makes for the magic of glass.

It also determined where the centers of glassmaking developed: all these natural ingredients had to be sourced, and the major glassmaking centers developed where the materials were available through local supplies or through easy trade (which also provided an export route). Venice is perhaps the prime example of the influence of supply and trade. Visit the city today and you are assaulted on every side by shop windows offering a dazzling, but commonly garish, display of glass. Today, many of these wares are no longer actually made in Venice, but for centuries it was the glassmaking capital of the Western world. In 1291, the already long-established guilds and manufacturers moved offshore, to the island of Murano. This not only freed the city from the threat of fire from the furnaces, but also allowed a cloak of industrial secrecy to settle around the jealously guarded tricks of the trade. The death penalty was imposed on tradesmen who leaked the island's secrets.

The source of their silica was an open secret, however. In northwestern Italy, the river Ticino flows out of the Alps past Milan. The Ticino carries, torn from the heart of the Alps and tumbled, ground, cleaned, and scrubbed by the river, quartz sand, gravel, and pebbles that are unusually pure—like those of the Coosa River of Alabama. It was the pebbles, the *cagoli,* that the glassmakers treasured.

In the middle of the fifteenth century, Angelo Barovier, one of the artisans of Murano, by carefully sourcing and purifying his ingredients, created a revolutionary product: *crystallo,* or crystal glass. It was the first truly colorless, transparent glass, and its optical properties were superior to anything that had been produced before. Transparent glass created an explosion of technology, fueling the scientific revolution of the Renaissance through microscopes, telescopes, and the laboratory. Transparency also vastly improved the quality of windows.

Today, when Venetians do make their own glass, they no longer use the *cagoli* of the Ticino River, but rather turn, like many glassmakers, to the forests outside Paris for their raw material. A visitor to Fontainebleau (where Napoleon abdicated—twice) might be drawn there by the magnificence of its château, but other treasures lie in its forest. Around thirty-five million years ago, a warm sea inundated much of northwestern Europe, and this sea retreated and returned over and over again. To the southeast, the Alps were still forming, rising from the forces of Africa crushing into old Europe. As from time immemorial, while the mountains rose, the elements chastised them for doing so, eating into the newly exposed rocks, eroding and destroying them. The *cagoli* headed south, but other debris from the Alps was carried northwestward by rivers to the encroaching sea, along the way grinding and sorting the sand that would be disgorged into the sea at the river's end. This sand was then caught up in the dynamic coastal processes we saw in chapter 5, all the time being cleaned and winnowed. As the sea made its final retreat, these sands were left stranded, and they are preserved today as the Fontainebleau sandstone. The rivers and the sea had done a fine job of cleaning the sand, but water later percolating through it leached out even more of the impurities, leaving huge tracts of sand that can be over 30 meters (100 ft) thick, fine, white, clean, and all of roughly same-sized grains—in other words, ideal for making glass. Fontainebleau has long been one of the premier glass sands in the world and today is a focus of major international glassmaking companies.

The important characteristics of a sand suitable for making glass are that it should have a high silica content and as few impurities as possible. Silica sand is rarely pure; the most common pollutant is iron, which, even in minute quantities, coats the grains, producing a variety of yellow and green colors in the glass (which can be reduced by adding manganese); even modern sheet glass often looks green from the side. For everyday glass, a typical proportion of silica sand grains would be 97 percent. For more sophisticated applications, such as for ophthalmic glass, the raw material must be 99.7 percent silica and contain less than 0.013 percent iron oxide.

In the United States, crumbling mountains and winnowed beaches allowed the first small glassmaking enterprise to be set up in Jamestown, Virginia, in 1608. It was established primarily to manufacture trading beads, but it soon fell victim to famine. Glassmaking didn't really take off until 1739, when Caspar Wistar (whose grandson we met in chapter 7, presenting a dinosaur bone for George Washington to examine) opened a factory in New Jersey, where there was an abundance of clean, ocean-washed sand, forests to fuel the furnaces and provide potash, and oyster shells to supply the calcium. The technology's secrets were still well guarded, and the British would not permit their own glass experts to emigrate. German glassblowers, brought over by Wistar, himself an immigrant from Germany, provided the expertise that began the American industry. And it is now a huge industry. The United States consumes well over eleven million tons of glass sand every year, the majority of it for everyday containers and plate glass, but also significant quantities for more sophisticated applications. Major glass manufacturers have tens of thousands of specialty products.

Alan Macfarlane and Gerry Martin have written, in *The Glass Bathyscaphe: How Glass Changed the World,* a compelling social history of glass, and they invite the reader to imagine a world without it. The evolution of the modern sciences would have been impossible without microscopes, telescopes, and laboratory vessels; as they write, "glass transformed humankind's relation with the natural world." How could Copernicus and Galileo have observed, imagined, and described as they did, and how could their theories have been tested, without glass? How could Antony van Leeuwenhoek have seen his miniature worlds? Passing an electric current through a wire and making it glow is simple, but it's not a lightbulb until it can be enclosed in a thin transparent glass container, strong enough to contain a vacuum and prevent the air from destroying the filament. Medicine, photography, cars, computers, television, navigation, laboratories, long-distance communication, mirrors, spectacles, the Hubble telescope, art, fine glasses for fine wine—the list is endless. As Cinderella's slipper and Alice's looking glass opened up new worlds for them, so does glass for us.

H

The *hourglass:* symbol of time, a satisfying conspiracy of sand, glass, and the physics of granular materials. There are records of "sandglasses" in ships' inventories from the fourteenth century, and it was the development of marine navigation that required them as a means of accurately measuring intervals of passing time. The technology of the marine compass seems to have arisen in Italy, quite possibly borrowed from much older Chinese instrumentation, and it is likely that Italian glass craftsmanship led to the design of an accurate sandglass. (One is depicted in Siena frescoes from 1338.) Until the needs of navigation, time had been of only relative concern and water clocks had served most purposes. In the often freezing European climate, however, water was hardly ideal, and relying, as such a device did, on a stable base for a consistent flow, it certainly would not work on board a ship. Sandglasses are essentially unaffected by heat, cold, and movement, and they were therefore ideal for nautical purposes, sometimes being referred to as "sea clocks."

Sea clock hourglasses came in two different versions, though neither of them ran for an hour. The thirty-second version measured the interval during which knots at measured intervals in a rope running out astern from the vessel were counted, thereby giving the speed—in knots. The thirty-minute version monitored the crew's shifts ("watches"), each one consisting of eight half-hours.

The glass needs to be made with extreme accuracy, and its slope should equal the angle of repose of the sand. The ratio between the size of the hole through which the sand flows and the size of the grains is critical—get it wrong and the sand will jam and not flow smoothly. In chapter 2, we saw how the pressure directly under a pile of sand is reduced through the formation of structures of arched chains of grains, and it is this behavior that allows a sandglass to work without clogging. Fine natural sand was often used but was not ideal; thus ground-up eggshells, metallic sand, and other materials were substituted. Complex sandglasses were made of several reservoirs so that fractions of a time interval could be measured. Hourglasses were, of course, superseded by the invention of accurate clocks. However, Queen Victoria is reported to have had one installed on the pulpit of her church: her sermon tolerance was said to have been twenty minutes. We still use them today, inefficiently, for timing board games and cooking eggs. They are collector's items—and their imagery is ancient and modern. There was a time when an hourglass was placed in a coffin as a (perhaps unnecessary) reminder that the sands of time had run out, and they have been popular symbols on gravestones. The classic opening of a long-running soap opera features an hourglass and the words, "Like sands through the hourglass, so are the Days of our Lives." And, as Jorge Luis Borges wrote in "Happiness," "Whoever looks at an hourglass sees the dissolution of an empire."

The largest hourglass in the world is at the Nima Sand Museum in Japan, built to celebrate the local "singing sands." It contains a ton of sand and is turned at midnight on December 31 each year.

I

Islands, artificial ones made of sand. Building artificial islands for a variety of purposes—agriculture, bridges, lighthouses—is nothing new, and sand has always been the most available material. Dredging operations need somewhere to deposit the spoils, and creating new land is an obvious solution. Balboa Island, off Newport Beach, California; Harbor Island, near Seattle; and the Venetian Islands in Biscayne Bay, Florida, were all created in this way. Harbor Island, designated as the largest artificial island in the world at the time of its construction in 1909, lost its title in 1939 when Treasure Island surfaced from the waters of San Francisco Bay to support the Golden Gate International Exposition.

Today, where to put dredged material has become more of a challenge. Every year, more than 200,000 dump truck loads of sand and mud need to be dredged from shipping channels to keep the port of Baltimore open for business, and the problem of where to put it has become acute. Reclaiming islands in the Chesapeake Bay that have suffered from erosion (see chapter 4) is one possibility; constructing an entirely new island in the bay is another. While the Chesapeake authorities tend to frown on such ideas, island-building projects are underway around the world. For example, two new islands are being planned off the coast of Israel—one for a new airport, the other for homes, businesses, and recreation. Understandably, there is considerable debate as to their impact on natural sand movement and changes to patterns of erosion and deposition. However, by far the most grandiose projects have been taking place in Dubai (the home of the world's only beach whose sand is temperature-controlled by a system of buried pipes).

Despite the admonishment of Matthew 7:26, there is, in principle, nothing wrong with building on sand—as the proportion of the world's population living on coasts demonstrates, both historically and today. From an engineering perspective, well-compacted, well-drained, level sand provides an excellent base. But building on sand that is vulnerable to liquefaction should be avoided, as should participation in nature's game of moving sand. Certainly, the powers-that-be in Dubai, for any number of reasons, have no inclination to heed the words of Matthew; what they are doing can only be described as mind-boggling.

In the shallow waters of the Persian Gulf, unimaginable volumes of sand are being dredged and formed into vast complexes of artificial islands, redefining the term *megaproject*. Three projects are nearing completion, one is underway, and the largest is still being planned. Their stated purpose is to solve Dubai's lack of shoreline—and to provide expensive new real estate. Three of the complexes are shaped like palm trees and will together add 520 kilometers (320 mi) of shoreline—many times the current length. The fourth is "The World" (the project itself has been named the eighth wonder thereof), an artificial archipelago of three hundred islands forming a map of the world and selling for an average of $30 million per island. The World covers an area about half the size of Manhattan and is built from enough sand (325 million cubic meters, or 425 million cubic yards) to make the concrete to build several hundred Pentagons. The next project, still on the drawing board, is reported to be named "The Universe"—although it will, modestly, depict only the solar system.

Dubai is situated on the edge of the Arabian Desert, and so sand supplies would not seem to be a problem. But desert sand, the grains rounded and smoothed by the wind, will not do. To create islands that cohere, the sand must be angular— marine sand. Gigantic dredgers rip sand from the sea floor of the Persian Gulf and "rainbow" it from huge hoses into position until, in The World, France, Green- land, California, and Los Angeles (these last two are separate islands) rise above sea level. The sand then suffers "vibro-compaction" in preparation for major con- struction projects. Dubai has now essentially exhausted its marine sand resource, with effects on the marine environment that have yet to be seen. The whole ar- chipelago is surrounded by a breakwater whose volumetric statistics are similarly biblical. This and the other breakwaters around the Palm complexes completely change water and sediment circulation. The construction of buildings and infra- structure on the islands is in its early stages, but the scale of what is planned— dozens of luxury high-rise hotels, offices, recreational facilities, and housing—will certainly test Matthew 7:26.

J

When the wicked fairy waves her wand, keep an eye on your *jewelry*—that family heirloom, the several-carat diamond mounted in gold, could vanish.

A few years ago, in the Democratic Republic of the Congo, a mine worker was digging in a narrow, deep pit in the sand, hauling bucketfuls to the surface, when out of the sand emerged a 265.82-carat diamond. The typical diamonds from the mine were less than a carat; this was a monster, and its ultimate value remains shrouded in the secrecy that still characterizes this often ethically dubious business. The di- amond had arrived there after a long journey from its birthplace in the immense pressures deep within the Earth, having been violently jetted upward through the crust, eroded, and tumbled along riverbeds to its resting place. As we have seen, the composition of sands betrays their origins, and if they came from the crum- bling of precious mineral deposits, then they will contain precious minerals. The rivers—and often waves and ocean currents—winnow the sands, concentrating grains according to their weight. The dark smears on beach ripples are formed this way, as were Thomas Edison's Long Island black sands (chapter 5)—and the Congo diamond deposits. These naturally concentrated sandy deposits of valuable minerals are called *placers* and are economically vital.

California was founded on placer sands: forty-niners during the Gold Rush sought the metal not only in subsurface mines, but in the streams and rivers that drained the gold-bearing ores. Panning the stream sediments was backbreaking work, and so a technological breakthrough was called for. It happened in the form of hydraulic mining: miners used high-powered water hoses to erode the valley sides. The gold, being heavier than the rest of the dirt, collected in sluices, and everything else drained away. It has been estimated that over a twenty-year period 750 million dump truck loads of sand, mud, and gravel were dumped into the Central Valley—with dire environmental consequences.

Placer mining is an old technology. An ingenious early method (probably used in ancient Egypt) employed a fleece bag, the woolen side facing inward: water and sediment were passed through the bag, and the heavy gold flakes became embedded in the wool, remaining behind when the bag was emptied of lighter sand and gravel. A similar method was still being used in the mountains of the Caucasus in the 1930s; it also explains Jason and his Golden Fleece.

Diamonds, rubies, sapphires, garnets, and gold are mined from placer sands in many parts of the world; in places like Namibia, beaches have been stripped to bedrock in the search for precious gemstones. These glamorous minerals are not, however, the only vital products of placer deposits. Concentrated iron-rich sands (richer versions of Edison's Long Island sand, or Petrus van Muschenbroek's "Magnetick-sand" [chapter 1]) are found worldwide and have long been used as the raw material for steel making. The finest-quality ceremonial Japanese swords have always been forged directly from *satestu,* iron sand. The iron and steel industry of New Zealand's North Island is based on iron-rich beach placers (whose titanium content initially presented a challenge to blast furnace technology). While the natural processes of concentration can create a health threat when the minerals are radioactive (as on some beaches in India), platinum, tungsten, titanium, tin, niobium, zirconium, and other vital elements are all sourced from placers (see also *x, y, and z*). To detail where we would be without these elements would take another book.

K

In 2004, a convoy of dump trucks crossed from South to North *Korea* and returned, completing a deeply symbolic journey: the first commercial trade across the land border since 1950. Their return cargo? Sand. Today, sand is imported nearly continuously into South Korea, mainly for cement production.

International trade in sand would hardly seem to constitute significant commerce—after all, everyone has sand—but it does, because often the right kind of sand is in the wrong place. Not only is Dubai's abundant desert sand too round for its artificial islands; it's not the right kind of sand to build its golf courses: that has to be imported. Saudi Arabia bans the export of silica sand and imports specialty products such as Australian garnet-rich placer sand for sandblasting. Sand is a major commodity in international trade; in some parts of the world, it is sufficiently valuable that cross-border disputes are common and sand smuggling is a thriving business.

The construction boom in China has stimulated massive illegal sand dredging, damaging river systems and increasing flood risk; beaches have all but disappeared from some Hong Kong islands. Singapore is a crowded, but hardly sand-rich, country, and so land reclamation projects are big business. For some time, Singapore has been at loggerheads with neighboring Indonesia over illegal, often nocturnal, dredging of Indonesian sand for smuggling into Singapore. One of Indonesia's islands close to the border has almost disappeared—creating potential implications as to where the border will be. Indonesia has banned sand exports, and its naval patrols enforce the ban. Singapore's access to Malaysian sand became the contentious bargaining chip in negotiations over a new bridge between the two countries.

Regrettably, illegal sand mining is a criminal, economic, and environmental problem in many parts of the world.

L

From beaches to golf courses, deserts to gardens, sand contributes significantly to our *leisure* activities—it provides the setting and material for having fun.

Let's begin with the beach. Simply lying on the beach or making a sand castle by hand is not enough for some; there has always been a major industry in beach equipment and paraphernalia. Buckets, spades, sand-carving tools, and shapers are the basics, but there are more sophisticated items. In 1903, the leading item of the Wolverine Supply and Manufacturing Company in Pittsburgh, Pennsylvania, was "Sandy Andy." One of a now classic series of brightly painted tin and steel toys, Sandy Andy was a tower from which a sand hopper filled a small truck that then rushed down, emptied itself, and rushed back up again. Production continued at least until the 1950s, but the early versions are collector's items. Since, by their very nature, they are often damaged by weathering, sand abrasion, and overuse, there is a debate whether, in terms of their value, demonstrable use or pristineness is to be treasured more. Toys where pouring sand is the driving mechanism are sold today—made, of course, from plastic.

Sand provided the surface for the aerial adventures of the Wright brothers, car racing, and land speed records, and today it lends its qualities to a variety of beach sports. Virtually every sport can be played on the beach—perhaps most popularly soccer and volleyball. Sand yachting, sailing, or boating events take place on beaches all over the world, in a variety of vehicles, in which the driver stands or sits while hurtling along under the power of the wind. Sand kiting involves attaching yourself to a large kite with oversized skates strapped to your feet. Sand skiing is exactly that—cross-country skiing on sand. The physical character of the sand is all-important, as is the nature of the snow in the original activity. The analogy between snow and sand has also resulted in the popular sport of sandboarding, which takes place down the slopes of sand dunes. Sandboarding has become a major international sport, the basis for schools, television shows, magazines, internet communities, and new technologies to overcome friction and abrasion.

And once the sands are opened to the internal combustion engine, it's a whole new and often bizarre world. Open a copy of *Sand Sports* magazine and the scale of these activities becomes apparent—roaring out of the pages are custom vehicles of every kind, from motorbikes and snowmobiles to drag racers and anything else on four wheels, many of which could come straight out of a (dystopian?) science-fiction film. Large tracts of the world's sand are scoured and redistributed by these sand sports on a daily basis.

The beach has also provided a medium for political advertising: during recent

French election campaigns, flip-flops were handed out, the soles of which imprinted in the sand, with every step, the initials of a political party. And we mustn't forget that sand was the stuff on which the bodybuilding empire of Charles Atlas was founded.

Sand toys are not limited to the beach or the sandbox, the beach in the garden. Sand art, sand sculpting, and sand magic toys (the latter often based on the physics of granular materials) are widely available, as are meditative "executive desk toys" of cascading colored sands.

Sporting surfaces from baseball to cricket rely on sand, and newly developed sand-based materials for horse-racing tracks and equestrian arenas are credited with reducing injury and loss of life. Simply mixed with crumbled rubber from old tires or as part of a manufactured fiber-reinforced product, sand is a key ingredient in such surfaces.

The thermal properties of sand provide for its role in cooking, from a buried pig at a Hawaiian luau to a Bedouin rabbit, from the Iroquois popping corn in heated sand to Chinese vendors roasting nuts in sand-filled woks. When the cooking is finished, the pot can be effectively cleaned with sand. The thermal approach is skipped altogether in Iceland, where shark meat is simply buried in sand for several months until it rots, after which it is dried; it is, apparently, an acquired taste.

Cooling, rather than heating, is the basis for the remarkable invention of Mohammed Bah Abba, a Nigerian teacher. He simply put one locally made terra-cotta pot inside another and filled the space between them with water and sand. Slow evaporation of the water, combined with the insulating properties of the sand, provided a cheap refrigerator requiring no electricity. Fresh vegetables could be kept for weeks rather than days. Abba's "pot-in-pot" is a stunning example of the simplest technology having a profound impact; the implications for the health and welfare of rural communities made this the winner of a *Time* magazine "Inventions of the Year" award in 2001. A pot sells for forty cents.

Abba's invention is possibly stretching the definition of *leisure,* but then so is the Marathon des Sables, perhaps the most extraordinary association of sand with sport. The world's most grueling footrace, the marathon covers, over six days, 240 kilometers (150 mi) across the Moroccan Sahara Desert. You need a few thousand dollars and a medical certificate to compete in this event, but in 2007 more than eight hundred men and women from all over the world entered (and most finished); two Moroccan brothers have dominated the event for ten years, typically finishing in under eighteen hours.

And, finally, if you have leisure time on your hands, try playing with the Falling Sand Game on the internet.

M

M is for *mummies, music,* and *morphing,* an eclectic combination, but each entertaining in its own way. The role of sand in mummification results from its character as an effective desiccant. Bodies buried simply in hot, dry sand are naturally dried out and mummified. "Ginger," whose body was wrapped in matting around 3200 B.C. and buried in the Egyptian sand, is a permanent resident of the British Museum and one of the oldest known mummies. Similar naturally desiccated mummies are found in Mexico, the Chinese Taklimakan Desert, and South America. A Chilean mummy is claimed to be much older than Ginger, and in Peru the naturally preserved remains of more than forty dogs have been excavated, buried in human cemeteries.

Sand effectively dries out a body but leaves the skin taut and brittle. It was to address this aesthetic problem that the ancient Egyptians developed the art of artificial mummification, using, among other materials, natron (soda from dried-up lake beds in the desert). It may have been for this purpose that Pliny's apocryphal traders were transporting natron when they invented glass. The Egyptians also used tar in the embalming process—the Arabic for "tar" is *mummiya,* but the term may have arisen through a misunderstanding, the dark, resinous-looking skin being mistaken for tar. Some of the tar used for mummification in Egypt has been identified (using biomarkers) as coming from the area of the Dead Sea (see the description of Sodom and Gomorrah in chapter 2) or from Gebel Zeit (which means "oil mountain" in Arabic), on the coast of the Gulf of Suez, close to modern oil and gas production (see *reservoir*); these natural tar occurrences commonly take the form of sand saturated with tar or bitumen. The wealthier the client, the less sand and more extravagant preservatives were used in Egyptian mummification: the poor had to rely on nature. Sand continued to play its desiccating role in the preservation of bodies in Italian monasteries (into the nineteenth century) and the preparation of shrunken heads, and is used today in laboratories and for drying flowers.

In *music,* sand is the sound producer in various rhythm instruments, such as rattles and sand blocks. Leroy Anderson, the composer of many widely recognized pieces, such as "Sleigh Ride," used sandpaper to evoke the sounds of soft-shoe dancers in his "Sandpaper Ballet." Anderson was imitating the likes of Fred Astaire performing the "sand shuffle" and the sand-dancing repertoire of the British comedian and entertainer Tommy Cooper. The sounds of sand, real or electronic, continue to be heard in various modern compositions. And Brian Wilson, of Beach Boys fame, found inspiration at his grand piano by placing it in a large sandbox installed in his living room.

Morphing, in the medium of animation, special effects, and computer graphics, exploits the fluid and shape-shifting character of sand. *The Sand Castle* (1977) is a delightful and whimsical work by Co Hoedeman, a Dutch-Canadian animator. The movie, which won an Academy Award for Best Short Animated Film, features the Sandman, a little character painstakingly constructed from wire, foam rubber, and sand, who emerges from the sand to create and sculpt a tribe of idiosyncratic sand creatures, each with its own function and shape to match. Together, they build their new home, the sand castle, the completion of which they celebrate enthusiastically— until the arrival of an uninvited guest, the wind.

And then, of course, there is the Sandman, one of the more sympathetic adversaries of Spider-Man from Marvel Comics and the third installment of the *Spider-Man* movies (2007), directed by Sam Raimi. The makers of the film immersed themselves in the physics of granular materials, developing algorithms to model flowing and blowing sand, shifting piles, and intergranular behaviors. In an interview for the Society of Digital Artists, Doug Bloom, the sand-effects supervisor, describes how they "decided the main idea was to give the sense that individual grains of sand have their own consciousness, and that they work together to form into a shape or to collapse." This Sandman is anything but whimsical: having found his body converted into sand after a bad encounter with a top-secret beach experiment, he can shape and re-form himself at will, turn his hands into sand weapons, merge with natural sand (from Arizona), and create particularly nasty sandstorms. The result is central to the movie and an extraordinary tribute to the art of special effects—and the dramatic character of sand.

N

Nanotechnology and *nanoengineering* are the modern sciences of making incredibly small things behave in unusual ways and achieve remarkable feats. We are only beginning to sense what is possible using submicroscopic materials, and silicon, derived, as we know, from sand, has a key role to play. There are one billion nanometers in a meter. A typical human hair is 80,000 nanometers in thickness, and nanotechnology deals with individual particles that are smaller than a human cell. A nano-sized piece of material behaves in ways radically different from the material's normal character, and this is the secret of many nanomaterials. For example, silica can be turned into an amorphous solid that contains minute holes, nanoscale pores, creating a huge internal surface area. The pores may be long and tube-like or like miniature cages, and into them and around them a variety of other molecules can be inserted. Such substances can create chemical tricks that previously seemed impossible, since they appear to run counter to the conventional laws of chemistry. A silica "aerogel" is the lightest (least dense) solid known, being 99.8 percent air. Silicon membranes have been created that, although strong, are only 15 nanometers thick and full of holes. These kinds of materials can serve as filters capable of removing individual molecules from polluted fluids; they are being used in biotechnology and as catalysts, for microchemistry, in ways that were once unthinkable.

Other applications are equally remarkable. Nanoparticles of silica referred to as "smart dust" can recognize specific molecules and offer the opportunity for robots the size of a grain of sand. Silica nanoparticles can provide the means of delivering cancer treatments, individual genes for gene therapy, and a host of other medical applications. A water-devouring silica nanofilm can keep your windshield clear without wipers. Nanoelectronics in computer chips, solar energy generation, and other applications promise step changes in size and efficiency. Nanotechnology also offers the possibility of inventing a nonpolluting concrete.

Medicine, biotechnology, electronics, construction, pollution monitoring and cleanup, smart fabrics, mineral extraction—the nanolist of applications goes on, and we have only just begun.

O

O is for *ostrich*—although, in spite of the common assumption and the adoption of the phrase into our language, ostriches *do not* bury their heads in the sand. The story goes that this myth originated, again, with the writings of Pliny the Elder, who described seeing an ostrich with its head entirely hidden in a bush. How this became translated into sand is unclear. Ostriches do, however, *eat* sand. They are occasionally observed with their heads on the ground ingesting pebbles and grit, and farmers will sometimes feed sand to the young birds. The grit is needed to assist with grinding food in their gizzards; indeed, many species of birds require sand or grit as a grinding supplement. It is often recommended that food put out during the winter for wild birds be sprinkled with a little sand.

Some worms and insects also consume sand—termites ingest sand from beneath the ground, carry it to the surface, and deposit it to build up their mounds. Analysis of termite hills has become a highly successful means of prospecting for mineral deposits, since the termites are sampling the geology at some depth. There is indeed gold in some of them thar hills.

People sometimes eat sand, a habit known as *geophagy.* Bizarre stories abound of such diets. An eighty-year-old woman in India is said to eat a kilogram of sand before breakfast. Japanese inhabitants of coral reef islands are said to have eliminated the need for doctors through eating the calcium-rich sand. All this is arguably patent nonsense, but *pica,* from the Latin for "magpie," is the medical term for an appetite for nonnutritional substances; it is a medical disorder.

And, on the subject of substance abuse, in 1934 *Ogden's,* a subsidiary of the Imperial Tobacco Company, issued a series of fifty cigarette cards on the theme of sand. Educational cards such as this, originating decades earlier in the United States as a means of putting to good use the card stiffener in a cigarette pack, became enormously popular. They are now collector's items; my own set of "The Story of Sand" cards covers, quaintly and more briefly, many of the topics of this book.

P

The renowned *porcelain* of Sèvres contained a number of unique ingredients that differentiated it from Oriental and Meissen porcelains. One of these ingredients was sand from Fontainebleau, the world-class glass sand. Porcelain and all ceramics require some form of fine sand to provide strength and thermal properties. Think of the heat-resistant tiles of the space shuttles: they are made from silica fibers, together with a ceramic binder, that provide an extraordinary ability to dissipate heat.

So when the fairy waves her wand, bid farewell to the family china—and your *personal care* and *pharmaceutical* products, and much of the contents of your kitchen cupboards too. We have already seen the role of fine silica as a filler in paint, but it also serves to thicken many gels, creams, and pastes—which covers most cosmetics. Together with other silicon compounds, it's a key ingredient in shampoos, conditioners, toothpaste, deodorants, nail polish, and so on. When added to a powder, silica products prevent caking and clogging and produce a well-behaved granular material. Pharmaceutical capsules, vitamin tablets, and powdered food products—cake mixes, flour, spices—take advantage of this.

Silica in various forms (particularly on a nanoscale) also plays a key role in the process of *papermaking.* The paper's surface character, particularly its absorbency, is often the result of a coating of a silica product—inkjet papers rely on this (and the ink's behavior is controlled by silica gel). Specialty silica and silicate papers refuse to burn at temperatures that melt copper. Less dramatically, cooks use silicone-impregnated parchment paper in baking. Unlike the petrochemical paraffin in waxed paper, silicone is both heat-resistant and inert.

Q

Quicksand, as we know, is unpleasant stuff and hardly qualifies as something useful. But the artificial and controlled production of a sort of quicksand has a number of valuable industrial applications. If air or another gas is forced upward through a bed of sand, the particles separate and become like a fluid—technically called a *fluidized bed.* Think of a hot-air popcorn-making machine: the uncooked kernels are forced into suspension and the heat is evenly delivered, resulting in uniform, rather than occasionally burnt, popcorn. It's the evenness that's the key—early fluidized-bed technology exploited it to fluidize particles of a catalyst that broke down heavy petroleum molecules into more useful light ones, the dynamics of the fluidized bed ensuring maximum efficiency of the catalyst. The method is used today in a variety of chemical processes, including the manufacture of polythene, making gas from coal, and cleaning up contaminated particles.

But perhaps the most important application of fluidized beds of sand is in incineration—materials burn more easily and efficiently in a fluidized bed. Not only do coal and other conventional fuels burn more efficiently, but so does garbage— "refuse-derived fuel," or RDF. Power generation that uses the technology creates fewer emissions and can be far more easily controlled than in traditional power stations. Fluidized-bed technology is also a way of generating hydrogen from methane for energy. Nanotechnology is increasingly teaming up with fluidized beds—most carbon nanotubes are created this way.

Perhaps you own an aquarium. If so, there's a good chance that you are using fluidized-bed technology every day and can watch it in action: many of the filter systems for cleaning the water are fluidized beds.

R

How many people in the world rely on water supplies flowing or pumped from underground? Controversial though the matter may be, how many of us rely on oil and gas flowing or pumped from underground? These critical resources do not occur in subsurface lakes, rivers, or caverns; they inhabit the microscopic holes in buried rock, very often the spaces between sand grains. Any rock that, like a sponge, is sufficiently porous to hold significant amounts of water or hydrocarbons is called a *reservoir.*

If rain falling in the hills soaks into porous and permeable rock layers that are tilted, sloping down below the surface out into the plains, the water will follow those layers until they flatten out. As long as there are no ways in which it can leak out, there it will collect, under the pressure of the overlying rocks, waiting in the pore spaces until a hole is drilled in the layer, releasing the pressure. The water will then flow upward, creating a water well. The layer through which the water moves and in which it collects is also known as an *aquifer,* "water carrier" in Latin. The water that we extract can be still flowing through the aquifer, continuously replenished, or if it has been slowly collecting there for a long period of time, it will not be replaced at the rate at which we remove it. Understanding and properly managing aquifers is critical, and we owe much of our ability to do so to Henry Philibert Gaspard Darcy, whose work on fluid flow revealed the physics of permeability and porosity (see *filter*).

Ninety-five percent of America's freshwater is underground, and a large proportion of it—30 percent of the water used for agriculture in the United States—is contained in one of the world's great aquifers, the Ogallala. The Ogallala, the main part of the High Plains aquifer, is formed from essentially unlithified sands, silts, and gravels that were carried off the eroding Rocky Mountains during the last twenty million years. Ancient sand dunes form parts of the aquifer, and the Nebraska Sand Hills are built on the Ogallala sediments. The reservoir is huge: it underlies an area of about 450,000 square kilometers (174,000 sq mi), covering parts of Colorado, Kansas, Nebraska, New Mexico, Oklahoma, South Dakota, Texas, and Wyoming. It is close to the surface, easily accessible, and 300 meters (1,000 ft) thick in parts, with an average thickness of 60 meters (200 ft).

The Ogallala contains staggering volumes of water—enough to fill Lake Huron—and supports the agricultural economy of the country. But its water is old, originally filling the reservoir during the ice ages, and it has been extracted at a rate many times faster than it is being replenished. Until the aquifer was understood and better managed toward the end of the last century, water levels had dropped on average 3 meters (10 ft) and, in some places, up to 30 meters (100 ft). We are still effectively "mining" the Ogallala water—it is a nonrenewable resource on our time scale. And the same is true of many of the world's aquifers, including those beneath the Sahara Desert, once thought to be constantly replenished, now understood to be filled with geologically ancient water.

The quality of an aquifer depends on its porosity and permeability—how much water it can contain and how efficiently it will give it up. These characteristics depend on the nature of the sand grains, the spaces between them, and the connections between the spaces. But however porous and permeable a reservoir may be, it will never give up *all* its water; the same surface tension effects that help build sand castles hold water onto and between the grains and keep *more water* in the rock than can be extracted. This same problem applies, very significantly, to oil reservoirs.

The world's largest accumulations of oil and gas are found in the spaces between sand grains. In Alaska, Saudi Arabia, Russia, and elsewhere, the reservoirs are made of sandstone—including carbonate oolite sandstones, which would seem to have little porosity (chapter 1). The reservoirs at the giant Prudhoe Bay field in Alaska are Permian and Triassic sandstones, deposited in shallow marine environments, deltas, and the beds of braided and meandering rivers. In the early days, it was thought that the sand grains and surface tension would allow only 40 percent of the oil in the reservoir to be extracted; technology has now increased that to almost 60 percent—but that's still a lot of oil left underground. In reservoirs that are particularly reluctant to give up their contents, sand is often forcibly pumped via the well bore and injected into the reservoir to open up space and increase the flow. Gas in between the grains is, of course, much less reluctant to leave, but recovery is still not 100 percent. (After they are emptied, the reservoirs can be used to store or sequester gas, notably carbon dioxide.) Most tricky of all are the tar sands of the Orinoco and of Canada (brought to the attention of the Hudson's Bay Company three hundred years ago by local tribes who used the tar to waterproof their canoes). Oil degraded by bacteria has formed the asphalt-like sticky stuff wrapped around the sand grains—and these are the largest oil accumulations in the world.

Reservoir engineering—the analysis, quantification, modeling, and prediction of the behavior of water and hydrocarbon reservoirs—is a whole science in itself, and it is critical for our way of life.

S

This is an excuse for a reminder, a celebration of the ways in which the word *sand* has contributed to our language. It appears in numerous place names—there are Sand Hills, Sand Points, Sand Creeks, Sand Lakes, Sand Banks, and Sand Rivers scattered across the map, never mind the places simply called Sand. Given that the word is the same in Danish, German, Norwegian, Swedish, and (with a *z*) Dutch, the list is long. We can eat sand tarts in the United States, *sandtortchen* in Germany, *zandkoekjes* in the Netherlands, and *sables* in France. Then there are all the living creatures and plants—sand grouse, cats, fleas, dollars, eels, shrimp, sharks, pipers, wasps, pears, pines, rats, and frogs, not to mention sand bubbler crabs. And people. Sandhogs were the workers who got their name in the 1880s during the construction of the Brooklyn Bridge and who still excavate the subterranean network beneath New York's streets and waterways ("if it's deeper than a grave, the sandhogs dug it"). Sandboys, proverbially happy, used to walk the streets selling "lily white sand" for cleaning and other uses; one of them fell tragically in love with "the Ratcatcher's Daughter," in an old Cockney ballad of the same name. Sandgroper, the colloquial name for a native of Western Australia, refers back to participants in that region's gold rush. We have sandbaggers in golf or poker, or any other sport; the Sandman, sometimes benign, sometimes sinister, bringing sleep and dreams; and the tirades of sandlot orators. Also in the sandlot, we have riots, constitutions, and baseball. Outside the National Baseball Hall of Fame and Museum in Cooperstown, New York (past which our sand grain journeyed in chapter 4), there is a compelling bronze statue of a boy in overalls and bare feet, hoisting a bat: "The Sand Lot Kid." Ever since a piece of sandy land in San Francisco in the late nineteenth century was set aside for "sandlot baseball," the term has come to embody the spirit and origins of the game to all of its aficionados.

Sand and grit serve as metaphors for strength of character, determination. Huck Finn says of Mary Jane: "In my opinion she had more sand in her than any girl I ever see; in my opinion she was just full of sand." And, speaking of determination, how many lines in the sand have been drawn through the course of history? Whether this phrase originated, as legend has it, at the Battle of the Alamo is open to debate.

To catalog the richness that sand has brought to the languages of the world is impossible. But one more example: in Malay, the word *desir* means the sound of sand blowing in the wind.

T

T is for *therapy*. Not only playing with sand, now a therapeutic tool for children and adults, but the benefits of being buried in it. Towering over the city of Beppu on Japan's Kyushu Island is the volcano of Tsurumi. The volcano has not erupted in over a thousand years, but its threat remains and its heat is palpable. Beppu is famed for its boiling pools of mud, hot springs, and other geothermal manifestations—including steaming pits of hot sand. For hundreds of years, the same family has been running the Takegawara Bathhouse, where one of the main attractions is being buried in hot black volcanic sand. Like gravediggers, the Takegawara sand ladies excavate your personal pit and bury you in steaming sand, for up to twenty minutes or so. Hot sand baths, *sunayu,* like this can be found in many places in Japan, in venerable bathhouses or on the beach.

Hot sand therapy is not peculiar to Japan. You can pay to be buried in sand in the Sahara and on the Red Sea coast, and witch doctors bury patients in Thailand. The "best" sand is often black, which in Japan means fragments of volcanic rock, and in Egypt means sand rich in iron minerals (placer deposits). Burial is reported (by, among others, Bernie, my local London taxi driver) to be good for rheumatism and arthritis as well as skin conditions, but there is no medical evidence for the therapeutic value of volcanic or iron sand. However, some manufacturers offer modern thermal sand beds with special sand for hospital use.

The story goes that certain Pacific Islanders used to bury their women in the sand with only their noses showing when foreign ships were sighted. Today a Polynesian spa offers sand massage and a facial scrub based on powdered local sand (see *abrasive*).

U

The story of *Ubar,* the "Atlantis of the Sands," is a reminder to us of how sand has preserved much of our global heritage that would otherwise have been lost. Ubar was a fabled trading city in the southern part of the Arabian Peninsula, home of the people of 'Ad, the descendants of Noah. According to the legend, the people of Ubar became decadent through the corruption of their fabulous wealth, and the city was destroyed in punishment and then buried in sand. In the 1980s, through a fortuitous series of events, Nicholas Clapp, a Los Angeles photographer, became fascinated by the legends and their possible basis in reality. After searching the desert by satellite and foot, Clapp and his team located what may well have been Ubar, largely buried in the sand. His book *The Road to Ubar: Finding the Atlantis of the Sands,* which documents his fascinating quest, has all the elements of a highly readable detective story.

Chapter 7 gave a sense of the fruitful collaboration between geology and archaeology. The examples are many, one of most recent and extraordinary being the work of a team of French and Syrian researchers at the four-thousand-year-old city of Al Rawda. On an aerial photo, the vague trace, a palimpsest, of a circular wall is visible in the sand. With geomagnetic imaging, a technique borrowed from mineral exploration that measures subtle changes in the magnetic field, a hauntingly clear image of a large city emerged, with concentric and radiating streets, buildings, and walls preserved beneath the sand.

Although the tomb of Ozymandias was described in 1799 by Napoleon's engineers in Egypt, it was not until the nineteenth century that the glories of the ancient Egyptian temples, carvings, and statuary of Thebes were liberated from the sand. Many of the temples and the "Ramesseum" (home of the "shatter'd visage" lying on the sand in Percy Bysshe Shelley's "Ozymandias") were built from sandstone. The Nubian Sandstone, around eighty million years old, was quarried for much of the building material from around 2000 B.C., and its engineering properties determined the methods and architecture. The Nubian today provides one of the world's great aquifers (see *reservoir*), which extends across much of northeastern Africa. Archaeologically, the Nubian Sandstone not only provided the material for Thebes but also composed the natural cliffs from which the monuments of Petra were carved and the rock within which the caves were excavated where the Dead Sea Scrolls were hidden.

It is not only the dryness of the climate that preserves ancient manuscripts—the desiccating properties of sand can do the same directly. On the other side of the Sahara from Thebes, Timbuktu holds dramatic illustrations of this. From around A.D. 1300 to 1500, the fabled city was a great seat of learning, with students and scholars coming from far away to study, learn, and debate. But after its fall, many of its archives became dispersed or lost. However, in recent years, following more peaceful times in Mali, literally thousands of manuscripts have been recovered from where they had been hidden, in caves or directly in the desert sand. Many are close to six hundred years old.

V

V is for *vines*, particularly those that produce grapes from which wine can be made—a process dear to my heart. Vines, at least the kind whose grapes make great wine, thrive on well-drained soils and deprivation. The relationship between wine and geology (the idea of *terroir,* or the influence of soil, rock, and climate on the character of a wine) is the subject of lively debate, but the fact that vine roots reach far below the surface for nourishment is undeniable. The vines of the Bordeaux region rely on the sediments of the Garonne and the Dordogne Rivers for their character. Many vines thrive in nothing but sand. Not far from the lighthouse at Aigues-Mortes (chapter 5), in the sands carried from the crumbling Alps by the Rhône

River, are *les vins des sables,* the sand wines. Organizations producing salt from the lagoons behind the Mediterranean coast planted vines in the sand dunes to make wine for the salt workers. After the devastating plague of phylloxera hit the vineyards of France in the late nineteenth century, it was noticed that the vines in the sand dunes were not affected: sand inhibits the lifestyle of the phylloxera-causing louse.

Vins des sables continue to be produced today, and they are very drinkable— but those in France are not alone. In the dunes on the coast of Portugal, not far from Lisbon, are the vineyards of Colares, planted in trenches to protect them from Atlantic gales. Increasingly rare because of pressure from coastal development, these vines also predate the ravages of phylloxera. So do the vines of the Kunsag region of Hungary, originally planted in the eighteenth century to stabilize sand dunes growing from the deteriorating soils of the plains of the Danube River. Similar survivors of phylloxera can be found in all of Europe's wine regions, wherever sandy soil protected an entire vineyard or even just a few rows.

Sandy vineyards in Australia also survived the blight, and some of California's oldest vineyards, in Contra Costa County, east of San Francisco, tell the same story. In what was once the state's primary wine-producing area, the Contra Costa vines, today hemmed in by development, are over a hundred years old, predating the phylloxera invasions. They are planted on sand dunes that formed as sea level dropped during the last ice age and wind went to work on the exposed sediments from the eroding Sierras. The vines produce first-class Zinfandels. Sand even enhances the already rich language of wine reviewers. In a 2007 description of the ten best Zinfandels in *Wine and Spirits* magazine, we read that, in one, "The flavors form a rhythm, like waves on a beach, first fruit, then chocolate, then a sand bar of tannin. Give this a year or two. . . ."

Although clearly the most widely appreciated, grape vines are far from the only plants to thrive in sand. Cranberries, for example, grow in bogs—sandy bogs. In 1816, on Cape Cod, Massachusetts, Henry Hall started up the first commercial cultivation of cranberries. He found that the biggest and juiciest cranberries came from where sand had blown from the coastal dunes over the vines. Layers of sand in a cranberry bog are a key ingredient today, and cranberry vines can be seen simply growing in the dunes. Cashew nuts and plum trees also grow in sand, as does pearl millet, one of the world's most resilient crops, a staple food for millions of people in semiarid regions. Mushrooms (for example, *Peziza ammophila,* "sand-loving") likewise can be found in sand dunes.

W

Sand is heavy stuff, and its **weight** makes for a variety of uses. The term *sandbagger* originates from the use, by gangs and other criminals, of cloth or leather bags filled with sand as effective offensive weapons (known as sand clubs, sand socks, or saps).

Sand has always been used as ballast in ships, to be dumped before a cargo is picked up, complicating—or perhaps aiding—any future forensic investigation. Sand is used to weight down essentially anything: basketball hoops, patio umbrellas, turf rollers, road signs, and road barriers. You can put bags of sand in the trunk of your car to gain extra traction on slippery winter roads (which may themselves have been sanded). In the days of hanging, the British government prescribed testing of the gallows using a sandbag of the same weight as the condemned man. Indiana Jones, in *Raiders of the Lost Ark,* switched the booby-trapped gold idol for a bag of sand—but he severely miscalculated the weight, and the trap was sprung. The ancient Chinese are reputed to have placed small sandbags on their eyes in an attempt to improve vision.

One of the latest methods of exercise and training is the "sacked session," which requires hauling bags of sand or pushing wheelbarrows full of sand around the gym. Or lie down, put a sandbag on your chest, and work until you are standing, holding the bag above your head. Less ambitious is working with small rubber sand balls (otherwise known as "therapy balls") about the size of a softball, filled with sand and sold in sets of two, one for each hand.

Specialist sand-filled leather bags are used by target shooters and engravers to provide sturdy but malleable support for rifle barrels or pieces of precious metal.

X, Y, and Z

Yes, the usual combination at the awkward end of the alphabet. Fortunately, *xenotime* is a mineral containing the element *yttrium* and is associated with *zircon* in placer sands (see *jewelry*). Yttrium is a clever element when used in metal alloys; it significantly changes their thermal properties, internal structure, and workability. Add it to glass and the latter becomes heat- and shock-resistant, useful for camera lenses, for example. It is used in lasers, in microwave filters, and as a catalyst. It is needed to stabilize synthetic cubic zirconia in fake diamonds. We have, of course, heard a fair amount about zircons (chapter 1 and above), and the long list of their uses would be a tedious way to end this chapter.

Far better, having got this far, to contemplate the calming influence of a *Zen garden* and cease worrying about what happens if the wicked fairy waves her wand—there wouldn't, after all, be a lot left to worry about.

10

Outward and Onward

Beyond Earth, beyond the Present

The whole universe may be as one plain, the distance between
planet and planet being only as the pores in a grain of sand,
and the spaces between system and system no greater than
the intervals between one grain and the grain adjacent.

Samuel Taylor Coleridge

But for a slice of luck, the whole world might have been made of sand.

Brian Aldiss, *Earthworks*

PROBING

It was the early hours of December 25, 2004. In England, Christmas stockings were being surreptitiously left next to sleeping children; in the United States, Christmas Eve celebrations were still continuing. Far, far away, a bundle of spectacular technology, looking like an inverted salad bowl set in a large, upside-down patio umbrella and wrapped in gold foil, slipped quietly away from its moorings and began its final journey. Twenty-one days into that journey, as it hurtled into an increasingly dense orange-brown haze, the technology awoke, its parachute deployed, the umbrella jettisoned. And it began filming a new world.

As the haze thinned, blurred outlines became visible, slowly resolving into a desolate, fractured landscape. Jagged ridges appeared, river valleys, coasts, dusty plains—and sand dunes. It would be satisfying to think that, somehow or another, Christiaan Huygens was watching the scene unfold; he would have been, like us, riveted, astonished. For this new world was Titan, the largest of Saturn's moons, first discovered by Huygens in 1655. It was in his honor that the bundle of technology, the *Huygens* probe, was named.

SCALE

As I suggested in the preface, because of the very nature of sand, this book is in many ways about scale, from the microscopic to, so far, the global. We have scrutinized sand in its many diverse roles—but only in the context of our planet and only in the present and the past. In this final chapter, we will broaden our horizons to look beyond the Earth and beyond the present. The first direction, as we will see, informs the second, and sand is there in both.

We met Huygens earlier in this book. As one of the luminaries in the 1600s, that extraordinary century of scientific inquiry and discovery, he moved in a broad but close community of illustrious colleagues. Huygens was a skilled card player and, with the encouragement of Blaise Pascal, wrote the first book on chance in gambling—or probability theory. He taught mathematics to Gottfried Leibniz, a man for whom scale was fundamental (chapter 1), was a family friend of René Descartes, visited Isaac Newton (with whom he disagreed on the theory of light), and worked on telescopes with the Dutch philosopher Benedict de Spinoza. As a glass grinder and maker of telescopes, Huygens was a friend of fellow Dutchman Antony van Leeuwenhoek, the microscope pioneer, who called Huygens "the great sky-inquirer." In a time when science was doing one of the things it does best—shattering the limitations of scale—Huygens was discovering new worlds in one direction, van Leeuwenhoek in the other. The latter discovered bacteria, described in one of his many letters to the Royal Society as so small that "even if 100 of these very wee animals lay stretched out one against another, they could not reach to the length of a coarse grain of sand." The former was revealing the enormity of the number of stars, so extensive that to count them "requires an immense Treasury, not of twenty or thirty Figures only . . . but of as many as there are Grains of Sand upon the shore. And yet who can say that even this number exceeds that of the fixt stars?" (*Cosmotheoros,* 1698). Oddly, even though Huygens was a notable mathematician, he makes no mention of Archimedes' approach to this challenge (chapter 3).

The discovery of Titan, called by Huygens simply *Saturni Luna,* was made possible by the telescope technology he had developed. He also identified separate rings around Saturn and was the first to see and describe the Orion Nebula. There are now thirty-five named moons of Saturn; further ones were identified not long after Titan by Giovanni Domenico Cassini, a French-Italian astronomer whose telescope was even longer than that of Huygens. The "mother ship" from which the *Huygens* probe detached itself that Christmas day was named in honor of Cassini.

CHILD OF HEAVEN AND EARTH

The world of Titan revealed seemingly miraculously by *Huygens* is in many ways starkly alien, in others hauntingly familiar. Titan deserves to be a planet—it's larger than Mercury and is the only satellite in the solar system to boast a proper, if exotic, atmosphere. In Greek mythology, the Titans, children of Uranus and Gaia, were powerful deities, and Saturn's moon is a powerful and dynamic world, in spite of its temperature.

As *Huygens* plunged through the atmosphere of Titan, the shock wave generated temperatures of 12,000°C (22,000°F), but by the time it landed, the surface temperature was about *minus* 180°C (-290°F). The probe owed its survival in these extremes to the properties of silica-based insulation and ceramics (chapter 9). *Huygens* settled on the surface and sent information homeward for an hour and a half before it succumbed to its environment. Peering out into its new world, *Huygens* had a limited, but dramatic, view. Figure 44 shows what it was looking at. (Scale is important here—the objects are cobbles and pebbles, not boulders.)

It is a strange, but at the same time recognizable, world. The area around the pebble in the foreground has been scoured, a phenomenon one can see on any riverbank on Earth after a flood subsides or on the rock-strewn surface of a desert hamada. The pebbles are rounded and set on a bed of what looks like sand. Extending a sensor, *Huygens* dug into it, and it had the consistency of sand. It *is* sand (because of its size), but sand unlike any we have encountered, or could encounter, in our own world. Titan's atmosphere is made up of over 98 percent nitrogen, the remainder being composed of methane (the simplest hydrocarbon), carbon dioxide, and complex carbon molecules. The sand grains and pebbles are made of rock-hard frozen hydrocarbon ice, and possibly some water ice. When *Huygens* landed, it set off a puff of gas, probably methane, from the sediment beneath it, melted by the probe.

FIGURE 44. Our first view of Titan's surface, transmitted by the *Huygens* probe. (Not a fine-quality image, but impressive, considering where it was taken.) (Image © ESA/NASA –University of Arizona)

The ingredients may be exotic, but the processes clearly are not. Not only did *Huygens* collect images of ridges and valleys as it descended, along with a scoured bank of sand and pebbles, but its mother ship, *Cassini,* has continued to observe and map the surface of Titan, using radar to penetrate the thick hydrocarbon smog of the atmosphere. The landscapes that it has revealed appear to have been sculpted by the same physical processes as those on Earth—sinuous river valleys and tributaries, eroded hills, volcanoes—but the material for the sculpture is not granite or sandstone or limestone, but frozen hydrocarbons.

How can the processes of erosion work at such frigid temperatures, when there is no water, the dominant sculptor on Earth? There is good evidence that it rains on Titan, which has storm clouds (larger and more menacing than ours) that gather periodically around the polar regions. The rain falls rarely, but when it does there is an almighty deluge, scouring the landscape with flash floods that carry vast volumes of sediment along with them, potentially rounding the pebbles and cobbles. However, the rain is not water, but toxic liquid hydrocarbons. Some methane is included, along with other materials pulled out of the atmosphere where the Sun has stewed the methane into more complex molecules, possibly forms of plastic. A portion of Titan's sand may even be made from these molecules, rained out of the atmosphere, the rest of it by more familiar processes of weathering and erosion. The pebbles in Figure 44 could be giant hailstones.

Large dark areas on Titan had originally been interpreted as seas of liquid methane, but *Cassini*'s radar gaze dispelled this view. The dark equatorial regions are filled with seas, but seas of sand dunes, extraterrestrial ergs (Figure 45). The dunes are long and linear, 100 meters (330 ft) high, extending over 1,500 kilometers (930 mi). They have been compared in appearance to satellite views of dunes in Namibia and Arabia.

Titan's dramatic atmosphere appears to circulate from the polar regions toward and around the equator, concentrating the seas of sand dunes. Winds on Earth are driven by solar heating and cooling, and it had been thought that because Titan is so far removed from the Sun's warmth, surface winds would be negligible. Titan's gravity is low, its atmosphere thick, and its sand relatively light, all of which make moving sand quite easy. But the dune fields demonstrate that it is also a world with significant winds. The origin of the winds, as we now understand, is the over-

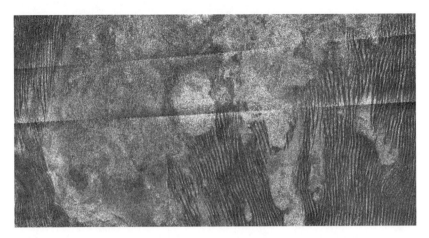

FIGURE 45. Dune fields on Titan. The area shown is approximately 200 kilometers (124 mi) across. (Image by NASA/JPL)

whelming presence of Saturn—the winds are caused by atmospheric tides as the gravitational embrace of the planet waxes and wanes, creating the effect of a kind of giant bellows. The resulting winds are still gentle by our standards—around half a meter per second (1 mph), but this is sufficient to move sand grains in Titan's environment, perhaps augmented by the periodic giant storms. By analogy with the formation of our own dunes, the linearity of those on Titan suggests variable wind directions; if we study the shapes of the dunes in Figure 45 as they wrap around the topography, we can begin to map wind directions on a moon over a billion kilometers away.

The mapping of Titan's surface is extraordinary: it reveals a ledger of familiar forms, written with a completely different chemistry.

WORLD OF THE HOURGLASS SEA

A geologist stands on a rippled bed of loose sand, carefully observing the layered sandstone cliff, documenting thicknesses, forms, and relationships. Moving closer, the observer examines the details of layering and architecture, notes the varying grain sizes with a magnifying lens, scratching the surface of the outcrop to get a better look. Routine field practice for a geologist—but this investigator is a robot, 380 million kilometers (240 million mi) from Earth. The robot is *Opportunity,* one of the two durable and observant Mars rovers that continue to trundle around, sending their detailed documentation back to the human geologists far away. The outcrop is Burns Cliff, in the wall of Endurance Crater, and the stories it tells are truly amazing. We shall return there shortly.

In 1659, when the Red Planet was in one of its close approaches to Earth, Christiaan Huygens took time off from perfecting the pendulum clock to turn his telescope toward Mars. He observed and sketched a large, dark, V-shaped mark on the planet's surface, noting its movement over a few days and calculating that the length of a day on Mars is the same as that of the Earth (the period is actually 24.6 hours). The dark feature that he watched, long referred to as the Hourglass Sea because of its shape, is Syrtis Major, a vast volcanic plateau. Huygens and others after him noticed that the appearance of the feature seemed to change with the seasons; with today's intimate knowledge of the surface of Mars, we now understand why. Figure 46 shows an area of Syrtis Major imaged by the Mars satellite *Odyssey;* the plateau is streaked with deposits of ever-shifting windblown sand accumulating in the lee of crater rims.

Unlike on Titan, the atmosphere of Mars is thin (pressure at the surface is less than 1 percent of that on Earth) and it's composed mainly of carbon dioxide, with a little nitrogen and argon and minute traces of other gases. But, also unlike Titan, Mars is exposed to the full effects of changes in solar radiation, its temperatures varying from comfortably above freezing in the summer to minus 140°C (-225°F) in the extremes of winter at the poles. This huge temperature range is more than sufficient to drive very strong winds that, in turn, are more than capable of moving the very fine grains of Martian sand. As today's research continues to show, if we simply change the terms for gravity, air density, and wind speed in Ralph Bagnold's equations, we can model the transport of sand *anywhere.* Indeed, in 1974, at the age of seventy-eight, Bagnold wrote a paper with Carl Sagan on aeolian transport on Mars.

In that paper, Bagnold and Sagan commented on the great Martian sandstorms of 1971, recorded by the *Mariner 9* probe. Sandstorms are regular stars of the Mars

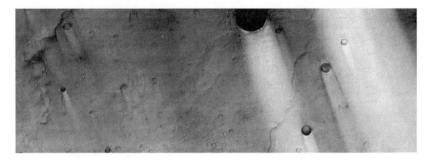

FIGURE 46. Sand streaks in Syrtis Major, on Mars. The area shown is approximately 60 kilometers (37 mi) across. (Image by NASA/JPL/ASU)

drama, as are dust devils on a scale unimaginable on Earth. These whirling dervishes of sand can reach heights many times those on Earth and cover huge areas of the surface at their base. Thorough studies of their puny earthly relatives are currently underway so that any attempt to put humans on Mars can take such things into account. We know very well the limitations to our understanding of the strange behaviors of granular materials on Earth and must significantly expand our thinking if we are to deal with them on Mars. For handle them we must. *Opportunity* was stuck for weeks in a sand ridge that was 30 centimeters (1 ft) high—to the rover, as to Bagnold in the Egyptian desert, the treacherously soft sand in front of it had looked no different from the rest. Human transport on Mars will require ingenuity—or, at least, gigantic wheels.

For any mission to send people to Mars, the availability of life-supporting resources will be a key issue. We shall look at what Martian sand is made of in more detail later, but much of it is derived from broken-down volcanic rock that, as on Earth, contains silicate minerals—a possible source of oxygen? But in order to "mine" and utilize sand, as well as to construct facilities, a knowledge of the granular behavior of the sand is critical. In a discussion of the kinds of Martian engineering questions that need answering, NASA asks whether "we understand granular processing well enough to do it on Mars" ("The Sands of Mars," 2005). The experts' answer: we don't yet know. Researchers at the University of Arizona, Cornell University, and the University of Colorado are designing innovative engineering approaches and methods of obtaining key data from the next generation of Mars missions. At the time of writing, one of these missions had just landed: the *Phoenix* is designed to dig trenches up to half a meter (20 in) deep, through the soil and, it is hoped, into Martian ice. It carries both atomic and high-powered optical microscopes to examine the materials from its excavation, including sand grains. Henry Clifton Sorby would have been delighted.

Phoenix is working in the north polar region of Mars to help answer critical questions, but it will not have the long life of *Opportunity*. Once the winter sets in, it will be entombed beneath the ice.

FROZEN DESERTS

The atmospheric circulation of Mars tends to move sand toward the poles, and it is in these regions that the sands of the planet become really spectacular. It is hardly surprising that, given the ample supplies of sand and the dynamic atmosphere, there are sand dunes on Mars. It is also not surprising that the rules of the aeolian game are the same as on Earth. Figure 47 shows dunes in the Syrtis Major region; if these features look familiar, compare them with the dunes in Plate 11. Barchan dunes are barchan dunes, whatever planet you find them on. This particular colony of dunes is marching toward the top left of the picture, the direction in which their steep, curving slip faces, their lee sides, are pointing. But when we look at the seas of dunes in the polar regions, something different seems to be happening; the asymmetry of the dunes tends to be less well developed, and evidence of significant migration is not easy to find.

The polar ergs of Mars are huge features—one sand sea is the size of Texas, and one dune is 475 meters (1,500 ft) high, the tallest known dune in the solar system. These sand seas occupy large topographical depressions (often craters) some 5 kilometers (3 mi) below the average surface elevation of the planet and are in places embraced by the towering walls of the polar ice caps. In the winter, as the ice sucks out more and more of the atmosphere, the ice caps expand and cover the dunes.

That the sand is seasonally buried by ice accounts for what would appear to be the violent harbingers of spring. As pressurized layers of ice and sand beneath the surface begin to melt, they break through the overlying ice, and jets of sand and

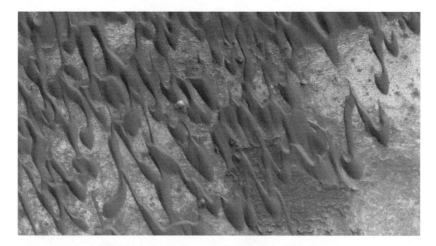

FIGURE 47. Barchan dunes in the equatorial region of Mars. The area shown is approximately 3 kilometers (2 mi) from top to bottom. (Image by NASA/JPL/Malin Space Systems)

carbon dioxide fire high into the atmosphere like giant alien geysers, showering the surface of the ice with falling sand.

The interplay between ice and sand creates extraordinary dune forms that differ from the typical denizens of Earth's great sand seas. Figure 48 shows some of the most studied dunes in the solar system—those in the Proctor Crater in the southern Martian Highlands. The image—a work of art in its own right—was taken in winter, and the bright areas are possibly ice or snow (carbon dioxide or water?)—or sand. The forms of the dunes are unusual. On a close look, they seem to have two slip faces, and some of the crests appear to have gullies cut in them, in contrast to the usual avalanche shapes. But they clearly are fresh dunes, with areas between that are sculpted by large ripples (left of the image). There is some evidence that the bright material moves backward and forward over the crests with the seasons and wind changes, but the appearance of the avalanches and the occasional gullies are odd. Elsewhere in the polar regions, the dunes have no obvious slip faces at all, appearing as domes of sand with rounded crests, along which gullies have clearly been eroded. But eroded by what? It seems highly likely that many of these dunes are, in a way, lithified, hardened and stabilized by frozen liquid between the

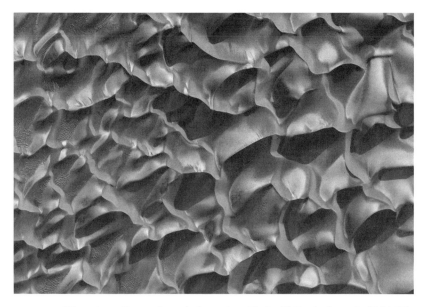

FIGURE 48. Winter dunes, Proctor Crater, in the southern Martian Highlands. (Image by NASA/JPL-Caltech/University of Arizona)

grains. Melting during the summer can cause slurry avalanches and the erosion of gullies. These dunes may be akin to those, saturated with water, that buried the Mongolian dinosaurs, as we saw in chapter 7, and are similar to dunes in Antarctica that seasonally freeze and thaw. It's also interesting to note that, perhaps analogously, in the days when Arabian dunes were slowed down by spraying them with oil, their slip faces disappeared; fresh sand continued to blow over them, but their shapes were domes rather than dunes.

The possibility that significant quantities of water, as well as frozen carbon dioxide, glue together the polar dunes of Mars is of enormous importance. The fieldwork that *Phoenix* will conduct is critical.

That Mars once had plentiful water, including oceans, is clear—and sand tells the story. Valleys contain meandering channels and point bars, deltas have formed at the mouths of canyons, and highlands show the hallmarks of water erosion. "Before" and "after" images show some signs of possible continuing, sporadic debris slides that could be liquid or dry granular flows. To see more of the testaments of sand to the early environment of the planet, we must return to the work of our robot geologist.

A FEELING WE'RE NOT IN KANSAS ANYMORE–OR ARE WE?

Plate 16 is a picture of an outcrop on Mars, Burns Cliff, assembled from many of *Opportunity's* carefully documented images. There is no question that we can apply a sort of planetary uniformitarianism to interpreting it. The cliff is built up of layers of sedimentary rock totaling around 7 meters (23 ft) in thickness, which can be divided into three general groups. The lowest section shows the massive internal features of sand dunes—cross-bedding—exactly as we see on Earth (left detail in the plate).

The middle section is made up of roughly parallel layers of sand and gravel. While the sands of Mars are typically composed of volcanic rock fragments, these are different. Yes, among the sands there are silicates, but many of the grains are formed of minerals that could only have been deposited in water. Among these are sulfate minerals, which would seem to make up much of the sand in Burns Cliff and elsewhere in this region of Mars. The sand dunes at White Sands National Monument in New Mexico are built of grains of gypsum (calcium sulfate), formed as desert lakes dried out and the wind picked up the crystals (chapter 1). Were there once dunes on Mars that looked like those at White Sands? Embedded in the sand are tiny spheres that have earned the affectionate name "blueberries," for they make the rock look like a fruit-studded muffin (right detail in the plate); they are made of the iron oxide mineral hematite. You have only to go to Zion National Park and look closely at the spectacular ancient dunes of the Navajo Sandstone there to see essentially the same things. In the Navajo Sandstone, they formed as part of the iron-rich diagenesis of the sandstone; on Mars, they seem to have formed in association with the evaporation of sulfate-rich water. The exotic mineral jarosite has also been identified, a watery sulfate of iron and potassium, found on our planet where waters are acidic.

The uppermost layers at Burns Cliff (the broken and jumbled rocks in the image) are different again. They contain small-scale cross-bedding and ripples, indications of a flowing current, and mud cracks, typical of periods when the sediment dries out.

We have ample evidence for the existence of water in the early days of Mars, but what would the scene have looked like? Kathleen Benison at Central Michigan University has put together a compelling story that relates the character of Burns Cliff to the environment 270 million years ago in Kansas and in modern acidic salt lakes in Australia. The situation in Kansas not long before the great Permian extinction began was unpleasant, to say the least. The rocks there record a desert environment, dunes surrounding ephemeral acidic salt lakes, the lakes drying out before being inundated by the next torrential rains. Such would seem to have been the conditions on Mars a long time ago—it probably hasn't rained there for a few billion years.

SLINGSHOTS AND ARROWS

All interplanetary probes rely on help in their journeys. The route to their destination is a complex one, as they swing around the solar system, employing the slingshot principle to steal energy from the gravity of planets to speed them on their way. *Cassini* reached Titan after two swings by Venus, a shot back past Earth, and a final boost from Jupiter. We shall continue our extraterrestrial sand quest in a kind of reverse slingshot: having looped around Mars, we shall now swing around Venus before returning to our home planet.

Venus was the destination of Russian choice in the early days of planetary expeditions, and, with some disasters along the way, it was the place where exploration records were set. In 1967, *Venera 4* was the first man-made object to enter another planet's atmosphere, and in 1970 *Venera 7* was the first to make a (relatively) soft landing and send information back home. Then on October 22, 1975, *Venera 9* sent us the first astonishing picture of a planet other than our own. It was a picture of rocks and sand.

Venera 9 was stalwart: it survived for nearly an hour on the truly evil surface of the planet. The composition of Venus's atmosphere is very similar to that of Mars—mainly carbon dioxide—but there's a lot more of it. The atmosphere is close to a hundred times as dense as ours, and the pressure on the surface would crush an automobile. The temperature is around 460°C (860°F), hot enough to melt lead, and one of the minor components of the atmosphere, sulfur dioxide, creates swirl-ing clouds of sulfuric acid. Images from recent missions whose radar has peered through the turbulent toxic soup of Venus's atmosphere show mountain peaks cov-ered in what looks like frost—obviously not frost as we know it, but perhaps metal snowflakes.

With an atmosphere this thick, it's probably difficult to stir up a good wind. But there is dust in the upper reaches, and Carl Sagan, following up on his work with Bagnold on Mars, evaluated the mechanics of Venusian aeolian transport in a paper in 1975. Radar images show wind streaks on the surface like those on Mars and features that can be interpreted as sand dunes—and the Russian images clearly show a sandy environment. Something is going on, but exactly how it all works re-mains to be seen. Venus is a strange place, and we shall return to it later as we reflect on our own planet.

As we turn homeward, we are reminded that we are not alone, and that the term *slingshot* is appropriate in another sense. Space is "full" of particles, many of them the size of sand, each on its own individual journey. *Full,* of course, is a relative term—what we encounter is hardly a cosmic sandstorm. But that there is debris flying around the solar system is no surprise; our own planet is a daily target. Four million extraterrestrial dump trucks shower their loads onto the Earth every year. Occasionally, a rather large chunk will have unfortunate consequences (as we saw in chapter 7), but most of the debris is sand and dust. The extraterrestrial compo-nents of sediments were first identified during the expedition of HMS *Challenger* in the late nineteenth century; tiny spherical grains, black and magnetic, were found on the ocean floor, with a character and composition unlike anything originating on Earth. Fine cosmic sand grains, *micrometeorites,* are found in polar ice (quite a few of them from Mars), and sand-sized debris from what may have been the largest collision in the history of the solar system is found entombed in 500-million-year-old limestones in Sweden. In the eastern United States, on the night of Novem-ber 12, 1833, many witnesses thought that the end of the world had arrived. The sky was filled with fireballs and flaming arrows—Native Americans described the event as "the night the stars fell." As was later understood, the Earth was on its annual encounter with the cosmic debris shed from the comet Tempel-Tuttle. This comet slingshots through the solar system every thirty-three years, renewing its cloud of rubbish, which, when the Earth encounters it in November every year, creates the Leonid meteor shower. The best performances are immediately after the comet it-self has visited, and 1833 was one such event: up to five thousand fiery flashes were to be seen every hour during the night. Much of the cloud is made up of cosmic sand grains, hitting Earth's atmosphere at speeds seventy times that of a bullet.

Comets are cosmic vacuum cleaners, collecting debris from the outer reaches of the solar system, but they are also very inefficient vacuum cleaners, spewing out clouds of material behind them. The chances of our encountering a comet on our journey back home are remote; if we do, the outcome could be catastrophic—but very interesting. Comets and their rubbish tell some extraordinary stories.

DEEP TIME, DEEP IMPACTS

In 1986, a veritable armada of spacecraft approached that most famous of comets, Halley's. The probe *Giotto*, on the first deep-space expedition of the European Space Agency (ESA), traveled in the company of a probe from NASA, two Russian craft, and two from Japan. The armada was in for a rough ride, but a productive one. The debris cloud around a comet, its *coma*, covers a tremendous volume of space; Halley's coma extends over 20 million kilometers (12 million mi) from the comet itself. It was, nevertheless, something of a surprise when one of the Japanese craft was hit by two particles the size of very coarse sand when it was 150,000 kilometers (90,000 mi) from the comet; fortunately, there was no serious damage. However, *Giotto*'s subsequent closer approach was compared by ESA's scientific program di-rector to a game of Russian roulette: "You may survive, but one shot will kill you." The analogy was accurate: as it got nearer to the comet, *Giotto* ran into a "wall of dust the size of grains of sand." The craft was knocked out of alignment, commu-nications were lost, and its camera was destroyed. But it did survive, and, along with the rest of the fleet, it sent back the first direct data from a comet.

Halley's nucleus is blacker than coal, but it's "fluffy"—around one-third the den-sity of water. It *is* made up of water (80 percent), plus dust, sand, and other debris, and some very interesting minor ingredients—carbon monoxide, carbon dioxide, and two other carbon compounds, methane and ammonia. But much of it is empty space.

To get a better look at what a rock is made of, a geologist will crack it open with a trusty hammer and look inside. This was the next important step in revealing the secrets of comets. On July 4, 2005, NASA's Deep Impact project dropped a battery-powered "impactor" onto an innocent, unremarkable, comet, Tempel 1, selected simply because of its proximity to Earth. The impactor's weight was equivalent to three heavyweight sumo wrestlers, and when it slammed into the small comet's nucleus it caused quite a stir—the event looked something like the impact sequence in Figure 13. Two hundred and fifty thousand tons of water and ten thousand dump truck loads of dust were ejected from a crater 100 meters wide and 30 meters deep (330 by 100 ft). The material that exploded from Tempel 1 is the stuff of the dawn of the solar system. Only the very coarsest particles reached the size of fine sand grains, most of them being finer dust; nonetheless, their composition has not only shed light on our system's origins but demonstrated that they are more complicated than once thought. In addition to water and gases similar to those around Halley, there were silicates, some primitive and unstructured, but some of them crystalline, including olivine, the same green mineral that populates some of Hawaii's famous beaches. The standard thinking about comets had been that, since they were formed in the outer reaches of the solar system, their material would be primitive, unprocessed by proximity to the cooking of the Sun. But this is clearly not the case—olivine and other silicates can be formed only by the solar stew. The solar system obviously has a much stronger circulation going on than had been thought.

It was well known that comets and asteroids contain organic material (compounds containing carbon, not necessarily produced by life), but Tempel 1 was a gold mine, a wide selection of complex organic molecules. These included methanol (CH_3OH), methyl cyanide (CH_3CN), acetylene (C_2H_2), and ethane (C_2H_6)—the kinds of compounds that can form the building blocks of life. Plentiful water and complex organic molecules were undoubtedly delivered to the early Earth by extraterrestrial bombardment. Tempel 1 also, incredibly, contained carbonate and clay minerals that generally require liquid water to form—where did they come from?

All this extraordinary information emerged from analysis of the effects of the impactor, without our ever laying hands on the stuff of comets. But that changed in January 2006, when a barbecue-sized device thudded into the Utah sand after a journey of more than 3 billion kilometers (1.9 billion mi). The Stardust project had sent a probe to encounter comet Wild 2, which was born beyond Neptune and is a recent visitor to the inner solar system. On the probe was aerogel, the lightest material known, made, through the wonders of nanotechnology, from silica (chapter 9). As the spacecraft flew through the comet's coma and particles slammed into it, the aerogel acted as a highly efficient shock absorber, capturing the particles and eventually returning them to Earth. This cornucopia of comet material, made up of very fine sand grains and smaller particles, continues to be analyzed by 150 researchers around the world, but some results are already clear. Crystalline olivine appeared again, showing, in the words of Scott Sandford of NASA's Ames Research Center, that "when the solar system formed, the solar nebula had to have been mixing like a son of a gun"—highly inconvenient for the traditional, simple theories of the origin of our solar system. Stardust's treasures also included polycyclic aromatic hydrocarbons, something of a molecular mouthful and called PAHs in the trade. PAHs are even more complex hydrocarbons, found in barbecue soot and automobile exhaust and as products of the incomplete combustion of materials such as wood, coal, and fat; their structure is made up of hexagonal rings, like a cross section of a honeycomb. They had been spotted before in interstellar space, but were not known to occur in comets. The PAHs from Stardust seem to be associated with oxygen and nitrogen—in terms of the way life might have started via incoming grains, the plot thickens.

A FLOATING RUBBLE PILE

Comets originate in the remote vastness of the solar system, in the Kuiper Belt beyond Mercury and the ominously named Oort Cloud far beyond Pluto. The Oort Cloud is a place of numerical excess—7.5 trillion kilometers (4.6 trillion mi) from the Sun and home to perhaps 10 trillion comets. Some of these icy, sandy, dusty, fluffy comets swing through the inner solar system and then disappear for sometimes thousands of years. *Asteroids,* on the other hand, come from closer to home, between Mars and Jupiter, originating from collisions early in the history of the solar system, a failed planet perhaps. The traditional distinction between asteroids and comets was their orbits and their composition, but the latter distinction is becoming less clear. Given their position closer to the Sun, asteroids

FIGURE 49. Itokawa, the floating rubble pile. (Image courtesy of Japan Aerospace Exploration Agency)

do not contain the ice of comets, but they are not necessarily solid rock either. In late 2005, the Japanese probe *Hayabusa* succeeded in actually *landing* on an asteroid, took pictures and samples, and took off again. In spite of engine problems, it will return to Earth in 2010. The asteroid in question, Itokawa, only a few hundred meters long, is a strange-looking character (Figure 49), resembling a dirt-covered potato in space. As documented by close-up images by the spacecraft, it's actually made up of dust, sand, gravel, boulders, and nothing—40 percent of it is nothing, in fact. It is simply a floating pile of rubble, held loosely together by its own gravity.

Did a huge colleague of this bizarre object wipe out the dinosaurs?

HOME: THE PALE BLUE DOT

In the quest for extraterrestrial sand, our journey has taken us vast distances. We may have occasionally strayed from the immediate material of interest, but it is always in the vicinity: sand and dust are the stuff the universe is made from. Peering into space, the Hubble telescope has seen "proto-planetary discs" in the Orion Nebula. Earthbound observatories are now confirming clouds of primordial sand, including silicates, swirling around young solar systems, slowly but surely clustering and coagulating into larger and larger pieces until planets are born. These views are being amplified by NASA's Space Infrared Telescope, "Spitzer," which, among other tasks, analyzed the results of the Deep Impact impactor's plunge. Spitzer is looking at already-formed planets and proto-planets light years away from Earth, detecting silicates and PAHs, watching births and deaths.

For the significance of scale from a human point of view, there is no image more iconic than that of our planet taken from the fringes of the solar system by the *Voyager 1* spacecraft on Valentine's Day 1990. Carl Sagan called the Earth in that image "a pale blue dot"; this became the title for one of his books and has been used ever since as a touchstone for reflection upon our home and our place in the universe. The "pale blue dot" provides a powerful image of scale, the minuteness of what is everything to us—and also a sense of fragility and vulnerability. We are fixated on fragility on our own scale, understandably, but as we look back over the 4.6-billion-year history of that pale blue dot, as we ponder its future and contemplate how our views are informed by our understanding of the universe, a broader perspective arises. One element of that perspective is how *different* our planet is— unique in our solar system, and arguably unique on a larger scale. Another element is how convenient for us that uniqueness is—but also how brief, in the grand scheme of things, that period of convenience really is.

And then as we begin to see home in more detail, we pass its sterile moon. We have much to be grateful to our moon for, but from the point of view of our topic, it holds little interest. Yes, there is sand on the Moon, but it has no life and is constantly pulverized over and over again by impacts in the absence of an atmosphere to protect and enliven it.

As we return to the embrace of our pale blue dot, it is its life that gives us the last element of perspective, not just life in the organic sense, but the life in its atmosphere and oceans, the vigor of its coasts and mountains, wet and arid lands, volcanoes and storms, and all the continuing, wonderful *activity.*

HITCHED TO EVERYTHING ELSE IN THE UNIVERSE

In 1911, in *My First Summer in the Sierra,* John Muir, the naturalist, geologist, and founder of the Sierra Club, wrote: "When we try to pick out anything by itself, we find it hitched to everything else in the Universe." I hope that this book illustrates that picking up a few grains of sand connects us not only to a particular beach or riverbank, but to the most recent journey that those grains have made, to the mountains from which they originated, to countless older journeys and cycles, and to the history of our planet. Muir was ahead of his time in many ways, this being one of them. It is only recently, perhaps facilitated by our views of the pale blue dot and other worlds, perhaps by radical cross-disciplinary conversations, that we have come to see the Earth as a *system*. We cannot, as Muir said, pick out the ocean, the atmosphere, or the continents and understand them independently, in the same way that we cannot look at a camshaft and understand how the internal combustion engine works. Reading the stories that a sand grain has to tell can only be accomplished through understanding the context, the system, in which that grain has played its role.

The uniquely dynamic character of our planet is a result of this system, the internal engine, the atmosphere, oceans, rocks, and living things constantly interacting in an immense, and immensely complicated, game. Scientists of all stripes attempt to understand the rules of this game, we hope with the same sense of wonder that Muir had. Sand has always been a major player in this game, not so much as a maker of the rules, but as the ground troops in it—the ones that live to tell the tale. As we have seen, sand bears witness to erosion and sea level change, landscape evolution and tectonics, the positions of continents, and climate—the vigor and character of the oceans and atmosphere—and the conditions preferred by life.

The complexity of the rules is nowhere better illustrated than in our climate. It is our climate that has enabled our existence and in whose turbulent embrace the future of our planet lies. Once again, in looking for the rules, we encounter the challenge of scale, not only the scale of the processes themselves, but the scale of change.

The words *climate* and *change* are inseparable. The Earth's climate has changed dramatically over the past 4.6 billion years, swinging between icehouse and hothouse conditions, cycling back and forth on slow and relatively rapid time scales. These changes occur on a scale of hundreds of millions and hundreds of thousands of years—when small changes happen on a scale of days, we call them "weather." On a human scale, we're in a warm period, but on a larger scale, this is simply another interglacial period of an ice age that began several million years ago.

The reason that forecasting (on whatever scale) is a challenge is simply the number of variables, the components that are interacting, and the fact that all *are* interacting as part of a complex system. The rules of the game are characterized by feedback, change causing change. Some feedback is positive, change augmenting change; some feedback is negative, change counteracting change. And all this is going on in a system that, at any given moment and any given place, is in a state of imbalance, ready to be tilted in one direction or another.

So far, we have seen the roles and testaments of our ground troops in the great game that is our Earth system today and in the past. What of the future? Feedbacks, positive and negative, conspire to regulate a system—if either type becomes dominant, then the condition of the system escalates. In the Earth system, we have two dominant components that are intimately linked regulators—plate tectonics and the composition of the atmosphere, both generals for the ground troops. We cannot help but notice that no other planet at the present time has plate tectonics or an atmosphere like ours—perhaps this can tell us something.

BACK TO THE FUTURE

It seems very possible that Venus, and perhaps Mars, had some form of plate tectonics way back in their history, that both had early atmospheres different from today's, and that something catastrophic happened to cause the change: crustal activity on both planets all but ceased, and their atmosphere was either largely lost (Mars) or turned into a toxic, escalating greenhouse (Venus). We also can't help but notice that our own planet seems to be fortuitously placed in what has been called the "Goldilocks zone"; for the entire diversity of life over the history of the planet, its setting has been "just right." Escaping from its early inferno, generously bequeathed water and complex organic molecules by comets and meteorites, relieved of a large portion of its lighter rocky materials by the formation of the Moon, Earth developed its dynamic tectonic and biological systems in a place that had just the right relationship with its star.

The single frame of the epic movie in which we feature today also happens to be "just right" for *Homo sapiens*—which is one reason our species has evolved the way it has. Today's environment is atypical—there are immense periods of the Earth's history that would have been intolerable for humans, and there will be a future that is equally so. At present, we are justifiably obsessed with one particular set of short-term feedbacks, ones that, as we shall see, have had a huge impact in the past, and will even more profoundly influence the Earth's future in the long term. Those feedbacks are, of course, between the atmosphere and surface temperature and, specifically, concern the role of carbon dioxide as a temperature regulator through the "greenhouse effect." An extraordinary fact is that the total amount of carbon dioxide in the Venus system is about the same as in the Earth system—the planets had similar origins. But Venus has suffered the nightmare of "runaway greenhouse," whereas we as yet have not. Why this crucial difference? The answer is that virtually all of the carbon dioxide of Venus has remained in its atmosphere, whereas on Earth most of it is locked away, or "sequestered," thanks to plate tectonics, the Earth system, and feedbacks. Soil contains twice the carbon that is currently in the atmosphere; the oceans contain fifty times as much. The Earth's crust holds, locked away over its history, 100,000 times the carbon that is in today's atmosphere.

In the (geologically) short term, sands will respond to a warming of the atmosphere in predictable ways. The physical games that sand plays will continue unchanged, but the pace of the games and their locations will shift. Rising sea level (as we have seen from the end of the last ice age episode) will move barrier islands and shorelines, rivers will adjust their profiles, and the balances between sand transport and deposition will adjust, both locally and globally. Migrating and vanishing beaches will be even more fickle in terms of supporting the tourism economy. Warming temperatures will change the levels of biological activity in the oceans and thus affect the formation of biogenic sands. Greater precipitation in some parts of the world will increase erosion rates and the sedimentary load of rivers—and the delivery of sand to changing littoral cells, whose sediment budgets will be revised. In other parts of the world, aridity will spread and relict ergs will be reactivated, increasing the transport of windblown sand. This, incidentally, might trigger its own example of feedback: if warming increases hurricane activity in the Atlantic, higher atmospheric levels of windblown sand and dust may, in turn, dampen the potency of those hurricanes—the jury is out (on both counts), because we don't yet understand the system fully.

But, as the world warms, other key feedbacks will continue in the background, as they always have, albeit on longer time scales than those in which we are currently and nervously interested. One feedback of global significance is weathering. As we saw in the first chapter, most sand is born from the decay and disintegration of rocks exposed to the ravages of the atmosphere. The chemistry of that process is vital and unique to our planet. The dominant mineral constituents of those rotting rocks are silicates, commonly combinations of calcium, sodium, and potassium with silicon and oxygen; formed at high temperatures and pressures within (and below) the crust, they are unstable at the surface. The carbon dioxide in the rain that falls on them and the moisture of the atmosphere that surrounds them are acidic, attacking these vulnerable minerals. When the carbon dioxide teams up with the calcium to form calcium carbonate—limestone—the silica is released to dissolve or re-form into clay minerals. Rivers carry the calcium (or in some cases, potassium or sodium) carbonate down to the oceans as their dissolved load. And on the shallow shelves of the world's oceans, life voraciously exploits those dissolved minerals to build shells. The carbon dioxide has been effectively removed from the atmosphere—potentially for very long periods of time as the shells fall to the sea floor and accumulate to build great thicknesses of limestone. Continually buried, those limestones will remain there until exhumed or subducted by the forces of plate tectonics.

This process, the carbon-silicate cycle, has been a crucial *negative* feedback mechanism and regulator. The warmer the surface of the Earth, the faster these chemical reactions of weathering take place and the more carbon dioxide is drawn down from the atmosphere—the greenhouse effect is reduced and the planet cools. Cooler conditions result in less vigorous weathering and less sequestering of carbon dioxide—and things start to warm up again. And it's not just temperature; the total *area* of rocks exposed to weathering is also critical. As plate tectonics steered India ponderously but dramatically into its collision with Asia over a long period around fifty million years ago, the towering Himalayas rose from the closing jaws of the vise. This monumental increase in fresh rock exposed to weathering reduced the carbon dioxide levels in the atmosphere to the point where the resulting cooling was the precursor of the ice ages. Current levels of carbon dioxide in our atmosphere are perhaps ten times less than they were when the dinosaurs were wiped out.

Temperature driving carbon dioxide levels driving temperature—a complex but elegant example of the mechanisms driving the Earth system. Except that, unfortunately, it's not that simple. We have neglected a couple of further factors and feedbacks—one negative, one positive. The first is life, on all scales: even simple bacteria living on desert sand grains have a significant effect on atmospheric chemistry. Vitally, we owe much of the historic stability, such as it is, in the level of atmospheric carbon dioxide to the emergence and activities of plants, which soak up the gas as part of photosynthesis and breathe oxygen into the atmosphere. And not only do plants directly soak up carbon dioxide, but their chemical activity modifies soil and increases rates of weathering, further drawing down atmospheric levels. By about 300 million years ago, Earth had become a fecund place for plants—great global forests developed, and these would form the raw material for the global coal beds after which the Carboniferous chapter of the Earth's history was named. But those forests caused carbon dioxide levels in the atmosphere to plummet, helping plunge the world into a major ice age. The increased ice coverage reflected more of the Sun's energy, another positive feedback for cold conditions. Did the dramatic reduction in weathering rates allow carbon dioxide levels to build slowly back up and eventually bring the ice age to a close? Did this greenhouse effect swing the Earth into the subsequent desert conditions of the Permian and Triassic, playing a part in the great Permian extinction? There are too many feedback mechanisms going on, too many possible smoking guns for the forensics to be definitive.

The other factor we have neglected is the temperature of the oceans. As we have seen, calcium carbonate, oddly, is more soluble in cold water than warm: warm waters have less capacity to store carbon dioxide than cold. So, as temperatures increase and weathering and the delivery of calcium carbonate to the oceans increases, warming ocean waters have to give up carbon dioxide, returning it to the atmosphere. This is a *positive* feedback, reinforcing the effectiveness of the greenhouse. Cycles, feedbacks, causes and effects, checks and balances, budgets—a complex system.

And it is a system that becomes more complicated the closer we look. The ice age at the end of the Carboniferous period was probably helped along by yet another factor—the position of the continents. Or "continent," for it was one of those times (the sequence of events has often repeated itself) when all the Earth's landmasses had gone into a huddle, having drifted together into the megacontinent of Pangea. The intimate interplay between land and oceanic and atmospheric circulation was dramatically affected; ocean currents were rerouted and bodies of warm and cold air repositioned. Furthermore, Pangea lay close to the South Pole, a landmass that created a perfect platform for thicknesses of ice to accumulate. Changes in weathering patterns resulting from a single landmass fed back into changes in the chemistry of the oceans. Plate tectonics, in a minor way through volcanic activity returning gases to the atmosphere and belching out radiation-reflecting dust and smoke, and in a major way through relocation of the continents, has a fundamental influence on climate. And plate tectonics, as long as it keeps operating, will continue to shape the evolution of our planet.

ENDGAMES

As we have seen, sand, the sawdust of plate tectonics and Earth's "grinding economy," is diagnostic of the growth and dynamics of the continents. The composition and character of sand have evolved over time as plate tectonic processes first got underway and have since developed. But why do we—uniquely and fortunately—live in a world of plate tectonics at all? The crust of Venus is nothing but a rigid, thick shell, a "stagnant lid" on the hot interior; molten material from the interior still occasionally breaks through to create volcanic eruptions, but it cannot drive plates. One of the fundamental differences between our dynamic planet and stagnant Venus is water. Water, absorbed into the minerals of the Earth's plates, significantly changes their physical properties. Water makes rocks less rigid; it weakens them and reduces their melting points. All this makes the slabs of the crust and upper mantle easier to bend, easier to subduct, easier to move. Subduction, in turn, carries water-rich minerals into the mantle, enabling the slow but enormously powerful driver of convection. Plate tectonics drives water around the Earth system, and water helps drive plate tectonics. For Venus, it remains a chicken-and-egg problem: did it lose its water because of the absence of plate tectonics, or did plate tectonics cease because of the loss of water?

Another difference between ourselves and Venus is our moon. Regardless of the precise sequence of events that formed the Moon, one effect was to remove a large proportion of the lighter components of Earth's early crust. This changed the makeup of our planet and, literally, *left room* for the continents to move about. Put the contents of the Moon back into the Earth and the surface of our planet would be gridlocked with continental crust.

As long as our planet has plate tectonics, its dynamism will continue—but for how long? In the longer term (and now we are talking about a geological time scale), a number of different routes to a couple of different endgames can be foreseen. A future extraterrestrial astrobiologist landing a probe on Earth would reconstruct some of that history from sand—yes, it will still be here.

In the immediate (geological) future, the next few thousand years or so, the ice ages that began a few million years ago will likely reassert themselves. Negative feedbacks to the atmospheric carbon dioxide levels could become dominant, resulting in things cooling off markedly. In the meantime, warmer oceans may increase the rate of atmospheric transport of moisture to the polar regions, increasing precipitation and the buildup of snow and ice. If we are around to witness this, will we, like Ozymandias, look on our works and despair as the ice grinds our cities back into sand?

Over the next thousands, or hundreds of thousands, of years, there are, of course, other factors beyond carbon dioxide levels that will in all likelihood help foster this renewed frigidity. Variations in the Earth's tilt and its rotation around the Sun have always contributed to cyclic fluctuations of temperature and climate, on a variety of time scales, and will continue to do so. There is also a suggestion that every few hundred thousand years our orbit may carry us through a particularly concentrated cloud of interplanetary debris, reducing the effect of the Sun's radiation.

In the still longer term, the influence of that radiation will become increasingly important. The Sun is brighter and warmer now than it was earlier in the Earth's history and, as we see through our telescopes has happened elsewhere, it will eventually flare up and burn out—in perhaps a further 3.5 billion years. But in the meantime, the Earth will have changed; of the various paths this change may take, the following is a likely one. Continuing solar warming will feed carbon dioxide drawdown until it reaches the point where land plants can no longer survive. Plants have already adapted their methods of photosynthesis to adjust to the decrease in carbon dioxide since the heady days of the dinosaurs, but there will be a limit to such adaptation. And that limit will mark a "tipping point" for the planet. As plants die off, the land surface will be exposed to rampant erosion; stable meandering rivers will disappear and be replaced by braided torrents, which will pour huge volumes of sand and mud into the oceans (as happened on a lesser scale as a result of the great Permian extinction). The ocean basins will fill with sand, and the land will be covered by it. The continents, swathed in windblown sand and dust, will wear down, and the ocean water, displaced by the volumes of sediment and swollen by the melted ice caps, will all but cover them. The Earth will become a water world, looking very much as it did a billion years ago.

What happens next will depend on a number of factors, particularly plate tectonics. If the engine of the Earth's internal radioactive heat diminishes, the point will be reached when there is no longer sufficient energy to drive the plates. Or if the increasing temperatures of the Sun drive most of the water of the oceans irrevocably into the atmosphere, the loss of its "lubricant" will also spell the inevitable end of plate movement. Or a feedback in the plate-tectonics game will itself bring things to a halt: through the activities of subduction and volcanism today, huge volumes of entirely new material are added each year to the continents—could continental growth reach the point where room for maneuver is so limited that the process shuts down?

There is yet another intriguing possibility for the future of plate tectonics. Adrian Lenardic at Rice University in Houston and Australian and Canadian colleagues investigate the metabolism of planets. They have recently reported that increasing surface temperatures would slow down the cooling rate of the interior, which would become hotter and therefore less viscous—to the point where it would lose the strength to move plates (this is what may have happened to Venus). Their work indicates that the effect would start to be noticeable when surface temperatures exceed 60°C (140°F); this is indeed hot, but the highest temperature so far recorded on our planet was in Libya in 1922—58°C (136°F). This is feedback, yet again: climate changes influence plate tectonics and the resulting changes modify the climate.

Regardless of which mechanisms cause it, the cessation of plate tectonics will herald the end of the world. Any hope of a thermostat, of regulating feedbacks, will be gone. Whatever feedbacks remain in the Earth's system will determine whether the future is simply a "moist greenhouse" or, like Venus, a "runaway greenhouse," as the Sun's increasing inferno cooks the surface of the planet.

SURVIVORS

The diversity of today's life-forms will be long gone, but chances are that life—as we are beginning to know it—will survive into some phases of the final endgame. It will do so in the sheltered worlds between grains of sand.

We revise our understanding of life today on a daily basis. As we expand our awareness of what constitutes the habitats of living things and the requirements, the fundamental nutrients, of life, we begin to question the very definition of life itself. Is it a valid assumption that life has to be based on water and carbon? Can exotic metabolisms formed from, for example, silicon, exist in environments very different from our own?

In chapter 2, we saw the extraordinary diversity of hitherto unknown microscopic organisms that inhabit the sand of an ordinary beach, communities with a diversity as great as, or greater than, that of any other environment on the planet. Among these creatures are the rotifers and the tardigrades. Rotifers thrive everywhere, from the poles to the tropics, and tardigrades are extraordinarily hardy creatures, capable, as we would never have thought possible, of surviving in extreme physical and chemical conditions. Tardigrades can withstand temperatures from -200 to +120°C (-330 to +250°F)—from the unimaginably cold to temperatures higher than the boiling point of water. Put a tardigrade under pressures hundreds of times that at the Earth's surface or place it in a vacuum, dowse it in carbon dioxide, carbon monoxide, nitrogen, and hydrogen sulfide, and it will survive. Under "normal" conditions, tardigrades are known to live for more than a hundred years, entering a state of suspended animation when necessary. But even tardigrades don't rank as genuine *extremophiles,* forms of life that *require* bizarre conditions. We are constantly discovering new microbes that thrive under extremes of pressure, temperature, acidity, alkalinity, salinity, toxicity, and radiation; creatures that have no need of oxygen and derive their energy from, among other nutrients, sulfur. Extremophiles are found in the super-hot springs of Yellowstone National Park, in the depths of the ocean trenches, in the volcanic vents on the mid-ocean ridges, in between the grains of sandstones in Antarctica, and at great depths in the Earth's crust. They give us a glimpse of our primitive past—and our primitive future.

Vulcano Island rears out of the Mediterranean to the north of Sicily. It was home to Aeolus, the Greek god of the winds, and today is home, as its name implies, to an active volcano. The geothermal heat provides, as in Japan, a thriving tourist business based around therapeutic mud baths and "hot water beaches." The temperature in parts of the sand in those beaches is close to the boiling point of water, and the grains are home to an extraordinary little extremophile, the delightfully named *Pyrococcus furiosus*. It is a simple spherical microbe that forms colonies around the sand grains. It thrives in boiling water (and "freezes" if the temperature drops), requires no oxygen, and exotically incorporates the metal tungsten into its metabolism. Its relatives are found in the deep-sea vents where the plates spreading apart produce superheated toxic environments and weird metabolisms.

As future conditions on our planet deteriorate, the microscopic spaces between sand grains will provide safe havens for life, almost to the end. It is not difficult to imagine the descendants of *Pyrococcus furiosus* representing the last of life on Earth—or their relatives thriving today deep in the sands of Mars. It is possible that life on Earth began in the spaces between grains of sand, and that this is also where it may end.

EPILOGUE

A Desert Mystery

The treasure of Tutankhamun, the Great Sand Sea, a meteorite impact, intrepid explorers, and a father's "desert things." I began this book with a personal story, and I shall round it off with another, one that brings together several of its themes and inspirations.

In *Lost Oasis: In Search of Paradise,* Rob Twigger recounts:

> One night my son wanted to appease me because of some annoyance he had caused. "Show me your desert things," he said. "Show me your crystals and stones." However tired and grumpy I might be, he knew how to revive me. I unwrapped my treasure from its newspaper roll. The chipped flint axes, the silica-glass arrowheads, the granite pestle, the ancient pottery shards I'd found in the Gilf, numerous fossils, petrified wood, the jawbone of a gazelle, date kernels so desiccated they were hard as stone, and a flake of obsidian, minutely worked ten thousand years ago into a blade that perfectly fitted my hand.

I shall be eternally grateful to Rob for taking me into the desert, and for also showing me his "desert things." I confess I succumbed to the addiction of the desert, having been fortunate enough to journey in the footsteps of Ralph Bagnold, walk in the Gilf Kebir (a plateau the size of Switzerland in the far southwest corner of Egypt), cross the Great Sand Sea—and assemble my own collection of desert things, among which are pieces of the silica glass to which Rob refers.

In 1922, the world caught its first glimpse of one of the most spectacular archaeological treasures ever revealed. Among the exquisite and glittering contents of Tutankhamun's tomb was a piece of jewelry—a pectoral, or necklace—its stunning design centered on a winged scarab (the dung beetle much loved by the ancient

Egyptians). The piece is a subtle spectrum of blues, reds, greens, and turquoise; it depicts lotus, papyrus, and lily flowers, cobras, and the eye of Horus, wrought from semiprecious carnelian, lapis lazuli, turquoise, and colored glass. But it is the scarab that grabs your attention. This scarab is carved from a glowing yellow-green gem-like material, long thought to be some sort of semiprecious stone—perhaps, as suggested by Howard Carter, the English archaeologist who located the tomb, chalcedony. It was not until more than seventy years after its discovery that an Italian mineralogist, Vincenzo de Michele, determined that it was made of glass. But it was not the traditional Egyptian glass that, as inlays or replicas of gems, decorates so much pharaonic jewelry. This glass was much, much older than the appearance of humans in Egypt.

Ten years after Carter's discovery, during the time when entrepreneurial expeditions were opening up, mapping, and documenting the Western Desert of Egypt and Libya, Patrick Clayton, a surveyor with the Egyptian Desert Survey, made an unusual discovery in the Great Sand Sea. Far to the southwest of Cairo, lapping across the border into Libya, is an immense area of desert for which the word "inhospitable" might have been coined, a seemingly endless sea of wave after wave of massive dunes. The car had only recently replaced the camel as a practical method of desert travel, and Clayton was painstakingly crossing the Great Sand Sea in a pragmatically modified Model A Ford. While there were many challenges to his progress, both mechanical and natural, Clayton hardly imagined that he would find himself driving over glass. But there it was, glittering jewel-like in the sun, chunks as large as a man's head, together with a myriad of smaller fragments. Some small pieces showed the telltale signs of prehistoric toolmaking, but most were smoothly irregular, opalescent, of an almost fluorescent, unearthly yellow-green color. This was the discovery of what came to be called "silica glass" or "Libyan desert glass," its origin a mystery and still the subject of intense scientific argument today. As de Michele, working with an Egyptian geologist, Aly Barakat, demonstrated, it was this glass that was used to make the scarab in Tutankhamun's necklace.

It's not surprising that glass should have been a common material in ancient Egypt—the basic raw material, sand, is hardly in short supply. But Libyan desert glass is different; not only is it old (around twenty-eight million years old, to be exact), but it is astonishingly pure. Its silica content is around 98 percent, making it far purer than early man-made glass, with a far higher melting point.

Natural glass is not uncommon, forming from rapid cooling of molten lava or, more rarely, from the fusion of quartz dune or beach sand by a lightning strike (creating the strange hollow tubes known as fulgurites). But this desert glass is different, not only in its overall purity, but in the chemical subtleties of its unique composition. Could the explanation for its formation be the violent impact and intense heat of a meteorite crashing into the Earth? The fundamental problem with this hypothesis is the absence of a smoking gun: where is the crater? There are impact craters in the Western Desert, but they are small, and it is unlikely that the energy generated by the impacts was sufficient to form the high-temperature glass—besides which, none of the strange desert glass is found anywhere near them. There have been larger candidates proposed for the crater related to the event, and anyone today can sit at home and browse satellite images of the desert, identifying large circular features. But on the ground, the evidence falls apart; circular patterns of hills and rocky outcrops can originate in a number of different ways, and meteorite impacts cause immense tracts of intensely shattered rock, evidence missing from the candidate impact sites and the area of the desert glass itself.

Other clues have a more direct bearing on this mystery. Twenty-eight million years ago, sedimentary rocks, now long eroded away, would have made up the landscape—sandstones such as the Nubian, which today forms the plateaus southwest of the dunes. The chemistry of this sandstone in the area of the Great Sand Sea is such that, if it were melted, the result would be the desert glass. This is a compelling piece of forensics, but the problem remains: what melted it? A possible answer comes from Siberia.

In the early morning of June 30, 1908, near the Tunguska River in northern Russia, the small local population of hunters and trappers witnessed a brilliant blue light splitting the sky. This was immediately followed by a barrage of explosions, a violent shaking of the ground, and a vicious, searing-hot wind that blew trees and the locals off their feet. Later investigation showed that sixty million trees had been scorched and flattened in a radial pattern over an area of 2,000 square kilometers (770 sq mi). It seemed to be the result of a meteorite impact, but there was no crater.

The best explanation for the Tunguska event is that an extraterrestrial object—in all likelihood a rubble pile like the asteroid Itokawa—exploded in the Earth's atmosphere high above the surface, causing a widespread thermal shock wave with the effect of several million tons of TNT being set off. This was undoubtedly a modest-sized object. What if it had been significantly larger? Could this be the explanation for the desert glass, the smoking gun in the crime story? It would certainly explain the lack of a crater, and though questions remain, it is the generally accepted origin for the glass.

In the spring of 2007, I fulfilled an ambition: I was walking across the rubble-strewn "street" between two of the great ridges of sand not far from where Clayton had encountered the glass. As geologists tend to, I walked slowly, my focus on the ground in front of me as I tried to visually tune in to the common rocks and pebbles so as better to spot something different. And then, catching the sunlight, there it was—my first fragment of desert glass. It was small but a treasure; I held it up, fascinated by the way it caught the light, by its intangible transparency and luminosity. The terrain was scoured by channels carved by the rare flash floods. Patches of slightly higher ground formed sandbanks, and I noticed that the particular patch on which I had made my discovery contained a concentration of glass fragments (Plate 17). I worked my way along it, discovering ever more stunning treasures. Having developed a search methodology, I would cross to another sandbank and, if I found nothing after a brief search, move on to the next, until I happened upon a rich horde. The larger pieces, polished and pitted by blowing sand, glowed translucently like jade; the smaller fragments, often faceted, sparkled transparently. I made a small collection of desert things, and then sat in wonder.

Back in Cairo, I went to the Egyptian Museum in quest of just one item from among the endless halls of treasures. I found it—to my surprise, an anonymous piece in a case of spectacular jewelry, with no hint of the story of the glowing scarab at its heart. I stayed for some time, enjoying what I felt to be a personal connection with Tutankhamun. On a couple of occasions, when visitors seemed to be looking specifically at the necklace, I took a small piece of desert glass from my pocket and told them the story. I didn't ask for baksheesh.

SOURCES AND FURTHER READING

References on sand in its various roles and manifestations are, like the material itself, a seemingly inexhaustible resource. Listed below, in alphabetical order chapter by chapter, are those sources that were most directly relevant to each section. With very few exceptions, each entry will appear only once, under the chapter where the subject matter is first referred to. Only those internet sites that are regarded as reliable and durable have been included. At the beginning, under "General," are key sources, available in print or via the internet, that were of use throughout the book. The author's comments appear in brackets.

General

Alden, Andrew. "Here's Sand in Your Eye." http://geology.about.com/od/sediment_soil/a/aboutsand.htm. [This site covers a wide range of geological topics well, with the nonspecialist in mind.]

Allen, Philip A. *Earth Surface Processes.* Malden, MA: Blackwell Science, 1997. [A classic treatment of the Earth as a system, integrating all scales from the global to the granular.]

Ayer, Jacques, Marco Bonifazi, and Jacques Lapaire. *Le sable: Secrets et beautés d'un monde minéral.* Neuchâtel: Musée d'histoire naturelle de Neuchâtel, 2002. [This lavishly illustrated paperback, published in tandem with a much-praised museum exhibition in Europe and Canada, is a treasure trove for arenophiles. It is, however, in French, and, regrettably, difficult to get hold of.]

European Space Agency (ESA). Home page. www.esa.int/esaCP/index.html. [The European portal to our planet, the solar system, and the universe; news, reports, and superb images.]

Greenberg, Gary. *A Grain of Sand: Nature's Secret Wonder.* St. Paul, MN: Voyageur Press, 2008. [For a visually stunning tour of the spectacular colors, shapes, and characters of sand grains, enjoy the specialist microscope photography in this book; also see the author's website, at www.sandgrains.com.]

Kuenen, Ph. H. *Sand: Its Origin, Transportation, Abrasion and Accumulation.* Annexure to vol. 62. Johannesburg: Geological Society of South Africa, 1959. [The classic discussion of the global nature of sand.]

Langer, William H. "Carved in Stone." www.aggman.com/articles/carved.htm. [A highly entertaining monthly column written by a U.S. Geological Survey geologist for *Aggregates Manager,* an online magazine serving the aggregates industry.]

Leeder, Mike. *Sedimentology and Sedimentary Basins: From Turbulence to Tectonics.* Malden, MA: Blackwell Publishing, 1999. [A university text, but an excellent resource for sedimentary matters on all scales.]

Logan, William Bryant. *Dirt: The Ecstatic Skin of the Earth.* New York: W. W. Norton, 1995. [A very readable account of soil, sand, and all the materials that are strewn over the surface of the Earth.]

National Aeronautics and Space Administration (NASA). Home page. www.nasa.gov. [The amazing portal to the solar system and satellite images of the Earth, all in the public domain.]

Okada, Hakuyu, and Alec J. Kenyon-Smith. *The Evolution of Clastic Sedimentology.* Edinburgh: Dunedin Academic Press, 2005. [An academic text, but a comprehensive review of the history of the subject and the personalities involved.]

Perry, Chris, and Kevin Taylor, eds. *Environmental Sedimentology.* Malden, MA: Blackwell Publishing, 2007. [Again, an academic book, but with excellent case studies of sediments and environmental issues.]

Pettijohn, F. J., Paul Edwin Potter, and Raymond Siever. *Sand and Sandstone.* 2nd ed. New York: Springer, 1987. [The classic text.]

Prothero, Donald R., and Fred Schwab. *Sedimentary Geology: An Introduction to Sedimentary Rocks and Stratigraphy.* 2nd ed. New York: W. H. Freeman, 2003. [Another good textbook, well illustrated and accessible.]

Siever, Raymond. *Sand.* New York: Scientific American Library, 1988. [A book for a general readership by one of my early mentors, with a focus on sandstones and their interpretation.]

Smithsonian National Museum of Natural History. *Geologic Time: The Story of a Changing Earth.* http://paleobiology.si.edu/geotime/main. [This should be the first port of call for anyone looking into our planet's history. Beautifully designed, interactive, and, of course, reliable.]

Summerfield, Michael A. *Global Geomorphology: An Introduction to the Study of Landforms.* Harlow, U.K.: Longman, 1991. [An excellent textbook, covering all scales of the Earth and its landscapes.]

U.S. Geological Survey (USGS). Home page. www.usgs.gov. [What can I say? Simply the most comprehensive portal to earth science, with, obviously, an emphasis on U.S. geology, but with a huge range of general articles, reports, images, teaching resources, and publications, all in the public domain. Plus, using the "Ask USGS" facility is a very effective and friendly way of getting answers to your questions.]

Preface

Eternal Sunshine of the Spotless Mind. DVD. Directed by Michel Gondry. Universal City, CA: Universal Studios, 2004.

Chapter 1. Individuals

Abbott, David M. "Investigating Mining Frauds." *Geotimes* 50 (January 2005): 30–32.

Bagnold, Ralph A. *The Physics of Blown Sand and Desert Dunes.* London: Methuen, 1954; reprint, Mineola, NY: Dover Publications, 2005.

Dunai, Tibor J., Gabriel A. Gonzáles López, and Joaquim Juez-Larré. "Oligocene-Miocene Age of Aridity in the Atacama Desert Revealed by Exposure Dating of Erosion-Sensitive Landforms." *Geology* 33 (April 2005): 321–24.

El-Sammak, A., and M. Tucker. "Ooids from Turkey and Egypt in the Eastern Mediterranean and a Love-Story of Antony and Cleopatra." *Facies* 46 (2002): 217–28.

Ford, Brian J. "The van Leeuwenhoek Specimens." *Notes and Records of the Royal Society of London* 36 (August 1981): 37–59.

German Artist Uncovers Australia. Videotape. Directed by Margaret Haselgrove. Sydney: Film Australia, 1995. [A visually stunning film of land artist Nikolaus Lang working in excavations of brilliantly colored sand.]

Gooding, Mel, and William Furlong. *Song of the Earth: European Artists and the Landscape.* London: Thames and Hudson, 2002. [Includes examples of the work of Nikolaus Lang.]

Greensfelder, Liese. "Subtleties of Sand Reveal How Mountains Crumble." *Science* 295 (January 11, 2002): 256–58.

Krinsley, David H., and John R. Marshall. "Sand Grain Textural Analysis: An Assessment." In *Clastic Particles: Scanning Electron Microscopy and Shape Analysis of Sedimentary and Volcanic Clasts.* Ed. John R. Marshall. New York: Van Nostrand Reinhold, 1987.

Mahaney, William C. *Atlas of Sand Grain Surface Textures and Applications.* Oxford: Oxford University Press, 2002.

Murray, Raymond C. "Collecting Crime Evidence from Earth." *Geotimes* 50 (January 2005): 18–22.

———. *Evidence from the Earth: Forensic Geology and Criminal Investigation.* Missoula, MT: · Mountain Press, 2004.

Pye, Kenneth, and Debra J. Croft, eds. *Forensic Geoscience: Principles, Techniques and Applications.* London: Geological Society, 2004.

Roberts, Dave. "The Nahoon Fossil Human and Animal Trackways: An Update." *QUARC Newsletter* 6 (June 1999). http://web.uct.ac.za/depts/quarc/newslet6.htm.

Rogers, J. David. "How Geologists Unraveled the Mystery of Japanese Vengeance Balloon Bombs in World War II." http://web.umr.edu/~rogersda/forensic_geology.

Sever, Megan. "Murder and Mud in the Shenandoah." *Geotimes* 50 (January 2005): 24–29.

Valley, John W. "A Cool Early Earth?" *Scientific American* (October 2005): 58–65. [Stories that zircons tell.]

van Leeuwenhoek, Antony. "Part of a Letter from Mr Anthony van Leuwenhoek, F.R.S. Concerning the Figures of Sand." *Philosophical Transactions of the Royal Society of London* 24 (1704–5): 1537–55.

van Muschenbroek, Petrus. "An Abstract of a Letter from Petrus van Muschenbroek, M.D.F.R.S. Professor of Mathematicks and Astronomy in the University of Utrecht, in Holland; to Dr. J. T. Desaguliers, F.R.S. Concerning Experiments Made on the Indian Magnetick-Sand." *Philosophical Transactions of the Royal Society of London* 38 (1733–34): 297–302.

Wentworth, Chester K. "A Scale of Grade and Class Terms for Clastic Sediments." *Journal of Geology* 30 (1922): 377–92.

Wilde, Simon A., et al. "Evidence from Detrital Zircons for the Existence of Continental Crust and Oceans on the Earth 4.4 Gyr Ago." *Nature* (January 11, 2001): 175–78.

Chapter 2. Tribes

Adam, Zachary. "Actinides and Life's Origins." *Astrobiology* (December 1, 2007): 852–72.

Alexander, A., F. J. Muzzio, and T. Shinbrot. "Effects of Scale and Inertia on Granular Banding Segregation." *Granular Matter* 5 (2004): 171–75.

Bagnold, Ralph A. *Sand, Wind, and War: Memoirs of a Desert Explorer.* Tucson: University of Arizona Press, 1990.

Bak, Per. *How Nature Works: The Science of Self-Organized Criticality.* Oxford: Oxford University Press, 1997. [Covers the wide-ranging implications of how sand piles behave.]

Ball, Philip. *The Self-Made Tapestry: Pattern Formation in Nature.* Oxford: Oxford University Press, 1999. [Contains a good review of granular materials.]

Brumfiel, Geoff. "Swirling Sand Mimics Liquid and Gas." *Physical Review Focus* (June 29, 2001). http://focus.aps.org/story/v7/st31#author.

Carson, Rachel. *The Edge of the Sea.* Boston: Houghton Mifflin, 1955; London: Penguin Books, 1999.

Chown, Marcus. "A Pattern Emerges." *New Scientist* (July 12, 1997): 34–36.

Experimental Nonlinear Physics Group, Department of Physics, University of Toronto. "What Is Nonlinear Physics?" www.physics.utoronto.ca/nonlinear/index.html. [Includes wide-ranging topics in granular physics.]

Fell, Andy. "Bacteria Could Steady Buildings against Earthquakes." *University of California–*

Davis News and Information (February 26, 2007). www.news.ucdavis.edu/search/news_detail.lasso?id=8040.

Hornbaker, D. J., et al. "What Keeps Sandcastles Standing?" *Nature* (June 19, 1997): 765.

Jaeger, Heinrich M., and Sidney R. Nagel. "Dynamics of Granular Material." *American Scientist* 85 (November 1997): 540–45.

Jaeger, Heinrich M., Troy Shinbrot, and Paul B. Umbanhowar. "Does the Granular Matter?" *Proceedings of the National Academy of Sciences* 97 (November 21, 2000): 12959–60.

Kestenbaum, David. "Sand Castles and Cocktail Nuts." *New Scientist* (May 24, 1997): 25.

Koppes, Steve. "Physicists Describe Strange New Fluid-like State of Matter." *University of Chicago Chronicle* 25 (January 5, 2006). http://chronicle.uchicago.edu/060105/physicists.shtml.

Lohse, Detlef, et al. "Impact on Soft Sand: Void Collapse and Jet Formation." *Physical Review Letters* 93/19 (November 5, 2004): article 198003.

Makse, Hernán A., et al. "Spontaneous Stratification in Granular Mixtures." *Nature* (March 27, 1997): 379–82.

Mikkelsen, René, et al. "Granular Eruptions: Void Collapse and Jet Formation." *Physics of Fluids* 14 (September 2002): 14.

Morell, Virginia. "Life on a Grain of Sand." *Discover* (April 1999). http://discovermagazine.com/1995/apr/lifeonagrainofsa491.

Morr, Megan. "When Flowing Grains Get Jam Packed." *Duke University News and Communications* (January 30, 2007). www.dukenews.duke.edu/2007/01/grainjam.html.

Nagel, Sidney. "Sand to Nuts." Transcript of interview by Alan Alda on "Life's Little Questions," Show 904 of the series *Scientific American Frontiers*, PBS (February 1999). www.pbs.org/saf/transcripts/transcript904.htm#6.

National Aeronautics and Space Administration (NASA). "Flowing Sand in Space." *Science@NASA* (November 17, 2000). http://science.nasa.gov/headlines/y2000/ast17nov_1.htm.

Neev, David, and K. O. Emery. *The Destruction of Sodom, Gomorrah, and Jericho: Geological, Climatological, and Archaeological Background*. New York: Oxford University Press, 1995.

Nowak, Sara, Azadeh Samadani, and Arshad Kudrolli. "Maximum Angle of Stability of a Wet Granular Pile." *Nature Physics* 1 (October 2005): 50–52.

Peterson, Ivars. "Digging into Sand: Sandpile Avalanches; Self-Organized Criticality." *Science News* 136 (July 15, 1989). http://findarticles.com/p/articles/mi_m1200/is_n3_v136/ai_7811593.

———. "Dry Sand, Wet Sand: Digging into the Physics of Sandpiles and Sand Castles." *Science News* 152 (September 20, 1997). www.sciencenews.org/pages/pdfs/data/1997/152-12/15212-16.pdf.

Schiffer, Peter. "Granular Physics: A Bridge to Sandpile Stability." *Nature Physics* 1 (October 2005): 21–22.

Shinbrot, Troy. "The Brazil Nut Effect—in Reverse." *Nature* (May 27, 2004): 352–53.

Umbanhowar, Paul B., Francisco Melo, and Harry L. Swinney. "Localized Excitations in a Vertically Vibrated Granular Layer." *Nature* (August 29, 1996): 793–96.

University of Chicago News Office. "Physicists See Similarities in Stream of Sand Grains, Exotic Plasma at Birth of Universe" (November 6, 2007). www-news.uchicago.edu/releases/07/071106.liquids.shtml.

Van Hecke, Martin. "Granular Matter: A Tale of Tails." *Nature* (June 23, 2005): 1041–42.

Weiss, Peter. "Mastering the Mixer: The Frustrating Physics of Cake Mix and Concrete." *Science News* 164 (July 26, 2003): 49–64.

Chapter 3. Sand and Imagination I

Archimedes (subject). "The Sand Reckoner." www.math.uwaterloo.ca/navigation/ideas/reckoner.shtml.

Borges, Jorge Luis. *The Book of Sand and Shakespeare's Memory.* Trans. Andrew Hurley. London: Penguin Books, 1998.

Canney, Maurice A. "The Use of Sand in Magic and Religion." *Man* 6 (January 1926): 13–17.

"Googol." *Time* 31 (February 28, 1938). www.time.com/time/magazine/article/0,9171,931101,00.html.

Kaplan, Robert, and Ellen Kaplan. *The Nothing That Is: A Natural History of Zero.* Oxford: Oxford University Press, 2000.

Kasner, Edward, and James Newman. *Mathematics and the Imagination.* New York: Simon and Schuster, 1940; Mineola, NY: Dover Publications, 2001.

Chapter 4. Societies on the Move

Brubaker, Jack. *Down the Susquehanna to the Chesapeake.* University Park, PA: Pennsylvania State University Press, 2002. [A delightful, detailed, and well-illustrated history, both social and natural, of the river.]

Dynesius, Mats, and Christer Nilsson. "Fragmentation and Flow Regulation of River Systems in the Northern Third of the World." *Science* 266 (November 4, 1994): 753–62.

Einstein, Albert. "The Cause of the Formation of Meanders in the Courses of Rivers and of the So-Called Baer's Law." *Die Naturwissenschaften* 14 (1926). www.ucalgary.ca/~kmuldrew/river.html.

George, Uwe. "Venezuela's Islands in Time." *National Geographic* 175 (May 1989): 525–61. [Discusses the geology and natural history of the tepuis.]

Hamer, Mick. "The Doomsday Wreck." *New Scientist* (August 21, 2004): 36–39. [On the USS *Richard Montgomery*.]

National Center for Earth-Surface Dynamics (NCED). "NCED Facilities: St. Anthony Falls Laboratory." www.nced.umn.edu/St_Anthony_Falls_Laboratory.html.

O'Driscoll, Patrick. "Scientists Watch Man-Made Flood of Grand Canyon." *USA Today* (November 28, 2004). www.usatoday.com/news/science/2004-11-28-grand-canyon_x.htm.

Schultz, Charles H., ed. *The Geology of Pennsylvania.* Harrisburg and Pittsburgh: Pennsylvania Geological Survey and Pittsburgh Geological Society, 1999.

Stokstad, Erik. "Geologic Modeling: Seeing a World in Grains of Sand." *Science* 287 (March 17, 2000): 1912–15.

U.S. Department of the Interior. *A Summary Report of Sediment Processes in Chesapeake Bay and Watershed,* ed. Michael Langland and Thomas Cronin. Water-Resources Investigations Report 03-4123, U.S. Geological Survey. New Cumberland, PA, 2003.

Chapter 5. Moving On

Abe, Kobo. *The Woman in the Dunes.* Trans. E. Dale Saunders. New York: Alfred A. Knopf, 1964; New York: Vintage International, 1991.

Ackerman, Jennifer. "Islands at the Edge." *National Geographic* 192 (August 1997): 2–31. [About the barrier islands of the eastern United States.]

Alexander, John, and James Lazell. *Ribbon of Sand: The Amazing Convergence of the Ocean and the Outer Banks.* Chapel Hill, NC: Algonquin Books, 1992; Chapel Hill: University of North Carolina Press, 2000.

British Geological Survey. *The Sea-Bed Sediments around the United Kingdom: Their Bathymetric and Physical Environment, Grain Size, Mineral Composition and Associated Bedforms,* by H. M. Pantin. Research Report SB/90/1. Keyworth, Nottingham, 1991.

California Coastal Commission. *The California Coastal Resource Guide.* Berkeley: University of California Press, 1987.

Carson, Rachel. *The Edge of the Sea.* Boston: Houghton Mifflin, 1955; London: Penguin Books, 1999.

———. *The Sea Around Us.* New York: Oxford University Press, 1951; 1991.

Coch, Nicholas K. "The Impending Coastal Crisis." *Geotimes* 53 (March 2008): 29–33.

Darwin, G. H. "On the Formation of Ripple-Mark in Sand." *Proceedings of the Royal Society of London* 36 (October 18, 1883): 18–43.

Davis, Richard A., Jr. *The Evolving Coast.* New York: Scientific American Library, 1996.

Dean, Cornelia. *Against the Tide: The Battle for America's Beaches.* New York: Columbia University Press, 1999.

de Villiers, Marq, and Sheila Hirtle. *Sable Island: The Strange Origins and Curious History of a Dune Adrift in the Atlantic.* New York: Walker Publishing, 2004.

Green, Amy. "Would It Still Be Miami Beach with Foreign Sand?" *Christian Science Monitor* (May 16, 2007). www.csmonitor.com/2007/0516/p01s03-ussc.html.

The House of Sand. DVD. Directed by Andrucha Waddington. Sony, 2005.

Lenček, Lena, and Gideon Bosker. *The Beach: The History of Paradise on Earth.* London: Pimlico, 1999.

Leonard, Jonathan Norton, et al. *Atlantic Beaches.* American Wilderness series. New York: Time-Life Books, 1972.

Lovett, Richard. "The Wave from Nowhere." *New Scientist* (February 24, 2007): 52–53. [On the 1929 Nova Scotia tsunami.]

Martin, Glen. "City's Beautiful but Hidden Sand Dunes." *San Francisco Chronicle* (July 20, 2006). http://sfgate.com/cgi-bin/article.cgi?f=/c/a/2006/07/20/MNGU1K2AV91.DTL.

National Geophysical Data Center (NGDC). Home page. www.ngdc.noaa.gov. [Like the U.S. Geological Survey, a superb resource for data on our planet, including images, teaching resources, and publications on natural hazards, the Earth's surface, and the oceans.]

National Oceanic and Atmospheric Administration (NOAA). Home page. www.noaa.gov. [The parent of the National Geophysical Data Center. Again, an extraordinary resource.]

National Oceanic and Atmospheric Administration, Coastal Services Center. "Beach Nourishment: A Guide for Local Government Officials—Geologic Regimes of the Atlantic and Gulf Coasts." www.csc.noaa.gov/beachnourishment/html/geo/index.htm.

National Park Service. "Cape Hatteras Lighthouse Relocation Articles and Images." www.nps.gov/archive/caha/lrp.htm.

"Ocean Beach Erosion and Natural Processes." www.sfenvironment.com/aboutus/openspaces/ocean_beach/erosion.htm.

Patsch, Kiki, and Gary Griggs. *Littoral Cells, Sand Budgets, and Beaches: Understanding California's Shoreline.* Institute of Marine Sciences, University of California, Santa Cruz, October 2006. www.dbw.ca.gov/CSMW/PDF/LittoralDrift.pdf.

Phillips, Angus. "Tall Order: Cape Hatteras Lighthouse Makes Tracks." *National Geographic* 197 (May 2000): 98–105.

Pilkey, Orrin H. "Beaches Awash with Politics." *Geotimes* 50 (July 2005): 38–39, 50.

Pilkey, Orrin H., Tracy Monegan Rice, and William J. Neal. *How to Read a North Carolina Beach: Bubble Holes, Barking Sands, and Rippled Runnels.* Chapel Hill: University of North Carolina Press, 2004.

Sackett, Russell, et al. *Edge of the Sea*. Planet Earth series. Alexandria, VA: Time-Life Books, 1983.

Scripps Institution of Oceanography. "Coastal Bluffs Provide More Sand to California Beaches than Previously Believed." *Scripps News* (October 12, 2005). http://scrippsnews .ucsd.edu/Releases/?releaseID=694.

Scripps Institution of Oceanography, Coastal Morphology Group, and Kavli Institute. *Living with Coastal Change: Informing Californians about Coastal Erosion*. http://coastalchange .ucsd.edu.

Sever, Megan. "Unlocking Blackbeard's Secrets." *Geotimes* 51 (October 2006): 34–37.

Shepard, Francis P. *Submarine Geology*. 3rd ed. New York: Harper and Row, 1973.

Squatriglia, Chuck. "Mystery of Vanishing Sand May Be Solved: Officials Think They Have Halted Erosion of Ocean Beach." *San Francisco Chronicle* (January 27, 2007). http://sfgate .com/cgi-bin/article.cgi?f=/c/a/2007/01/27/BAG8ANQ1OA1.DTL&hw=ocean+beach& sn=013&sc=513.

U.S. Department of the Interior. *An Overview of Coastal Land Loss: With Emphasis on the Southeastern United States,* by Robert A. Morton. Open File Report 03-337, U.S. Geological Survey. http://pubs.usgs.gov/of/2003/ofo3-337/index.html.

U.S. Geological Survey (USGS). "Coastal Change Hazards: Hurricanes and Extreme Storms." http://coastal.er.usgs.gov/hurricanes.

———. "San Francisco Bight Coastal Processes Study." http://walrus.wr.usgs.gov/coastal_ processes/pubs.html.

Woman of the Dunes. Videotape. Directed by Hiroshi Teshigahara. Argos Films and the British Film Institute, 1964. [Released in the United States as *Woman in the Dunes*.]

Woods Hole Oceanographic Institution. *Oceanus*. www.whoi.edu/oceanus/index.do.

Wright, Orville. Letter to his sister. October 14, 1900. www.smithsonianeducation.org/ educators/lesson_plans/wright/kitty_hawk.html.

Chapter 6. Blowing in the Wind

Bagnold, Ralph A. *Libyan Sands: Travel in a Dead World*. London: Hodder and Stoughton, 1935.

———. *The Physics of Blown Sand and Desert Dunes*. London: Methuen, 1954; reprint, Mineola, NY: Dover Publications, 2005.

———. *Sand, Wind, and War: Memoirs of a Desert Explorer*. Tucson: University of Arizona Press, 1990.

Chalmers, Matthew. "The Troubled Song of the Sand Dunes." *Physics World* (November 1, 2006). http://physicsworld.com/cws/article/print/26278.

de Villiers, Marq, and Sheila Hirtle. *Sahara: A Natural History.* New York: Walker Publishing, 2002.

Douady, S., et al. "The Song of the Dunes as a Self-Synchronized Instrument." *arXiv* 3 (January 28, 2006): 1–9. http://arxiv.org/vc/nlin/papers/0412/0412047v1.pdf.

The English Patient. DVD. Directed by Anthony Minghella. Miramax, 1996.

Gluckman, Ron. "The Desert Storm." *Asiaweek* 26 (October 13, 2000). www-cgi.cnn.com/ASIANOW/asiaweek/magazine/2000/1013/is.china.html. [On desertification in China.]

Hogan, Jenny. "Dunes Alive with the Sand of Music." *New Scientist* (December 18, 2004): 8.

Hunt, Melany L. Home page. www.me.caltech.edu/hunt. [Includes summaries of Hunt's research into granular materials and singing dunes.]

"The Itjaritjari." *New Scientist* (May 29, 2004): 53.

Jasper, David. *The Sacred Desert: Religion, Literature, Art, and Culture.* Malden, MA: Blackwell Publishing, 2004.

Kok, Jasper F., and Nilton O. Renno. "Electrostatics in Wind-Blown Sand." *Physical Review Letters* 100/014501 (January 11, 2008): 1–4. http://sitemaker.umich.edu/jasperkok/files/kokrennoprl2008.pdf.

Livingstone, Ian, and Andrew Warren. *Aeolian Geomorphology: An Introduction.* Harlow, U.K.: Longman, 1996.

McNamee, Gregory, ed. *The Sierra Club Desert Reader: A Literary Companion.* San Francisco: Sierra Club Books, 1995.

Merali, Zeeya. "Dune Tunes . . . The Greatest Hits." *New Scientist* (September 17, 2005): 11.

National Aeronautics and Space Administration (NASA). "City-Swallowing Sand Dunes." *Science@NASA* (December 6, 2002). http://science.nasa.gov/headlines/y2002/06dec_dunes.htm.

National Park Service. "Great Sand Dunes National Park and Preserve, Colorado." www.nps.gov/grsa.

Oldfield, Sara. *Deserts: The Living Drylands.* London: New Holland, 2004.

Saltation and electric fields (subject). "Electric Sand Findings Could Lead to Better Climate Models." www.ns.umich.edu/htdocs/releases/story.php?id=6256.

Schwämmle, Veit, and Hans J. Herrmann. "Geomorphology: Solitary Wave Behavior of Sand Dunes." *Nature* (December 11, 2003): 619–20.

Seife, Charles. "The Soft Footfall That Signals Dinner." *New Scientist* (April 3, 1999): 6. [On sand scorpions.]

Swift, Jeremy, et al. *The Sahara.* World's Wild Places series. Amsterdam: Time-Life Books, 1975.

Trexler, Dennis T., and Wilton N. Melhorn. "Singing and Booming Sand Dunes of California and Nevada." *California Geology* 39 (July 1986): 147–52. www.schweich.com/sbdA.html.

Twigger, Robert. *Lost Oasis: In Search of Paradise.* London: Weidenfeld and Nicolson, 2007.

Walker, A. S. *Deserts: Geology and Resources.* Online edition: http://pubs.usgs.gov/gip/deserts.

Waltham, Tony. "Pinnacles and Barchans in the Egyptian Desert." *Geology Today* 17 (May–June 2001): 101–104.

Webster, Donovan. "Alashan: China's Unknown Gobi." *National Geographic* 201 (January 2002): 48–75.

Williams, Caroline. "A Land Turned to Dust." *New Scientist* (June 4, 2005): 38–41. [On desertification in China.]

Youlin, Yang, Victor Squires, and Lu Qi, eds. *Global Alarm: Dust and Sandstorms from the World's Drylands.* Online publication. United Nations, August 2001. www.unccd.int/publicinfo/duststorms/parto-eng.pdf.

Zandonella, Catherine. "Shifting Sands." *New Scientist* (June 28, 2003): 40–43. [On the work of Jean Meunier in Mauritania.]

Chapter 7. Witness

Carson, Rachel. *The Sea Around Us.* New York: Oxford University Press, 1951; 1991.

Cross, Timothy A., and Peter W. Homewood. "Amanz Gressly's Role in Founding Modern Stratigraphy." *Geological Society of America Bulletin* 109 (December 1997): 1617–30.

Dickinson, William R. *Temper Sands in Prehistoric Oceanian Pottery: Geotectonics, Sedimentology, Petrography, Provenance.* Special Paper 406. Boulder, CO: Geological Society of America, 2006.

DiGregorio, Barry E. "Doubts on Dinosaurs: Yucatán Impact Crater May Have Occurred before the Dinosaurs Went Extinct." *Scientific American* (May 16, 2005). www.sciam.com/article.cfm?articleID=0000D7BD-7720-1264-B1DB83414B7F0000.

Dingus, Lowell, and David Loope. "Death in the Dunes." *Natural History* 109 (July–August 2000): 50–55. [On the Gobi Desert dinosaurs.]

Dudley Stamp, L. *The Earth's Crust.* London: George G. Harrap, 1951.

Eriksson, K. A., and E. L. Simpson. "Quantifying the Oldest Tidal Record: The 3.2 Ga Moodies Group, Barberton Greenstone Belt, South Africa." *Geology* 28 (September 2000): 831–34.

Fichter, Lynn S. *The Geological Evolution of Virginia and the Mid-Atlantic Region.* http://csmres.jmu.edu/geollab/vageol/vahist/index.html.

Fortey, Richard. *Life: An Unauthorised Biography.* London: Harper Collins, 1997. [An immensely readable account of the history of life on Earth by a leading paleontologist and science writer.]

Friend, P. F., and B. P. J. Williams, eds. *New Perspectives on the Old Red Sandstone.* Special Publication 180. London: Geological Society of London, 2000.

Grady, M. M., et al., eds. *Meteorites: Flux with Time and Impact Effects.* London: Geological Society, 1998.

Gutscher, Marc-André. "Destruction of Atlantis by a Great Earthquake and Tsunami? A Geological Analysis of the Spartel Bank Hypothesis." *Geology* 33 (August 2005): 685–88.

Hallam, A. *Great Geological Controversies.* 2nd ed. Oxford: Oxford University Press, 1989. [A review of the development of geological thinking: time and processes, the age of the Earth, continental drift, and mass extinctions.]

Holling, Holling Clancy. *Paddle-to-the-Sea.* Boston: Houghton Mifflin, 1941.

Koeberl, C., and K. G. MacLeod, eds. *Catastrophic Events and Mass Extinctions: Impacts and Beyond.* Boulder, CO: Geological Society of America, 2002.

Kraft, John C., et al. "Harbor Areas at Ancient Troy: Sedimentology and Geomorphology Complement Homer's *Iliad.*" *Geology* 31 (February 2003): 163–66.

Letourneau, Peter M., and Paul E. Olsen, eds. *The Great Rift Valleys of Pangea in Eastern North America: Sedimentology, Stratigraphy, and Paleontology.* Vol. 2. New York: Columbia University Press, 2003.

Mazumder, Rajat, and Makoto Arima. "Tidal Rhythmites and Their Implications." *Earth-Science Reviews* 69 (2005): 79–95.

Monastersky, R. "The Moon's Tug Stretches Out the Day." *Science News* 150 (July 6, 1996). www.sciencenews.org/pages/sn_arch/7_6_96/fob1.htm.

Nelson, Stephen A., and Suzanne F. Leclair. "Katrina's Unique Splay Deposits in a New Orleans Neighborhood." *GSA Today* 16 (September 2006): 4–10.

Nield, Ted. *Supercontinent: Ten Billion Years in the Life of Our Planet.* Cambridge, MA: Harvard University Press, 2007. [A lively account of past (and future) supercontinents.]

Paris, R., et al. "Coastal Sedimentation Associated with the December 26, 2004 Tsunami in Lhok Nga, West Banda Aceh (Sumatra, Indonesia)." *Marine Geology* 238 (2007): 93–106.

Retallack, Gregory J., et al. "Middle-Late Permian Mass Extinction on Land." *GSA Bulletin* 118 (November–December 2006): 1398–1411.

Rowe, Clinton M., et al. "Inconsistencies between Pangean Reconstructions and Basic Climate Controls." *Science* 318 (November 23, 2007): 1284–86.

Smith, David. "Tsunami: A Research Perspective." *Geology Today* 21 (March–April 2005): 64–68.

Travis, John. "Hunting Prehistoric Hurricanes: Storm-Tossed Sand Offers a Record of Ancient Cyclones." *Science News* 157 (May 20, 2000). www.sciencenews.org/articles/20000520/bob9.asp.

Vita-Finzi, Claudio. *The Mediterranean Valleys: Geological Changes in Historical Times.* Cambridge, U.K.: Cambridge University Press, 1969.

Ward, Peter. "Precambrian Strikes Back." *New Scientist* (February 9, 2008): 40–43. [A good review of biomarkers—extinctions and other stories.]

Ward, Peter D., David R. Montgomery, and Roger Smith. "Altered River Morphology in South Africa Related to the Permian-Triassic Extinction." *Science* 289 (September 8, 2000): 1740–43.

White, Rosalind V. "Earth's Biggest 'Whodunnit': Unravelling the Clues in the Case of the End-Permian Mass Extinction." *Philosophical Transactions of the Royal Society of London* 360 (2002): 2963–85.

Wong, Kate. "Lucy's Baby: An Extraordinary New Human Fossil Comes to Light." *Scientific American* (September 20, 2006). www.sciam.com/article.cfm?articleID=00076C1D-62D1 -1511-A2D183414B7F0000&sc=I100322.

Wood, Bernard. "Palaeoanthropology: A Precious Little Bundle." *Nature* (September 21, 2006): 278–81. [The science of the "Dikika baby" story.]

Chapter 8. Sand and Imagination II

Anderson, Lee. "Legends in Sand: The Evolution of the Modern Navajo Sandpainting." www.americana.net/sandpaintings_article.html.

Australian Aboriginal sand art (subject). "Art in Balgo." www.loreoftheland.com.au/ indigenous/kutjungka/index.html.

Critchley, Jay. "Beige Motel 2007." http://jaycritchley.com/beige.html.

Demaine, Erik D., Martin L. Demaine, Perouz Taslakian, and Godfried T. Toussaint. "Sand Drawings and Gaussian Graphs." *Journal of Mathematics and the Arts* 1:2 (June 2007): 125–32.

Denevan, Jim. Home page. www.jimdenevan.com.

Gerdes, Paulus. *Geometry from Africa: Mathematical and Educational Explorations.* Washington, DC: Mathematical Association of America, 1999.

Hébert, Jean-Pierre. "Sand as Medium: The Tradition." http://hebert.kitp.ucsb.edu/sand/ tradition.html.

Iowa Geological Survey. "From Sandstone to Sand Paintings." www.igsb.uiowa.edu/Browse/ Pikes%20Peak/sand_art.htm. [On Andrew Clemens.]

Kamat, Jyotsna. "Rangoli: The Painted Prayers of India." www.kamat.com/kalranga/art/ rangoli.htm.

Peterson, Ivars. "Sand Drawings and Mirror Curves." *Science News* 160 (September 22, 2001). www.sciencenews.org/articles/20010922/mathtrek.asp. [On *sona* sand designs.]

Rischmueller, Marian Carroll. "McGregor Sand Artist." *The Palimpsest* 26 (May 1945). http://clipclop.tripod.com/andrew/dox/sandartist.html. [On Andrew Clemens.]

Sardon, Mariano. "Books of Sand." www.marianosardon.com.ar/libros_arena_eng.htm.

Shapiro, Bruce. "The Art of Motion Control." www.taomc.com/home.htm.

Vanuatu Cultural Centre. "Sand Drawing." www.vanuatuculture.org/sand/index.shtml.

Vesna, Victoria, and James Grimzewski. "Nanomandala." http://nano.arts.ucla.edu/mandala/mandala.php.

Wheel of Time. DVD. Directed by Werner Herzog. Wellspring, 2003. [Film of the making of a Tibetan sand mandala.]

Wigan, Willard. "The World's Smallest and Most Wondrous Works of Art." www.willard-wigan.com.

Chapter 9. Servant

British Geological Survey, Minerals UK, Centre for Sustainable Mineral Development. Home page. www.bgs.ac.uk/mineralsuk/home.html.

Carr, Donald D., ed. *Industrial Minerals and Rocks.* 6th ed. Littleton, CO: Society for Mining, Metallurgy, and Exploration, 1994.

Clapp, Nicholas. *The Road to Ubar: Finding the Atlantis of the Sands.* Boston: Houghton Mifflin, 1998.

"Concrete Possibilities." *Economist* (September 23, 2006): 28, 30.

Corning Museum of Glass. "A Resource on Glass." www.cmog.org/index.asp?pageId=426.

Edwards, Richard, and Keith Atkinson. *Ore Deposit Geology and Its Influence on Mineral Exploration.* London: Chapman and Hall, 1986.

Falling Sand Game. www.fallingsandgame.com/sand.

Goho, Alexandra. "Concrete Nation: Bright Future for Ancient Material." *Science News* 167 (January 1, 2005). www.sciencenews.org/articles/20050101/bob9.asp.

Gui Qing, Koh. "Singapore Finds It Hard to Expand without Sand." *Reuters* (April 14, 2005). www.singapore-window.org/sw05/050414re.htm.

Hammer, Joshua. "The Treasures of Timbuktu." *Smithsonian* 37 (December 2006): 46–57.

Lovett, Richard. "The Day the Gold Rush Stopped." *New Scientist* (November 12, 2005): 60–61. [Review of the damage caused by hydraulic mining and its role in the roots of the American environmental movement.]

Macfarlane, Alan, and Gerry Martin. *The Glass Bathyscaphe: How Glass Changed the World.* London: Profile Books, 2003.

McGuire, V. L. "Water-Level Changes in the High Plains Aquifer, Predevelopment to 2003 and 2002 to 2003." Fact Sheet 2004-3097, U.S. Geological Survey. http://pubs.usgs.gov/fs/2004/3097.

National Aeronautics and Space Administration (NASA). "Aerogel: Mystifying Blue Smoke." http://stardust.jpl.nasa.gov/aerogel_factsheet.pdf.

Port Washington Public Library. *Sand and City.* www.pwpl.org/localhistory/sandmine. [An extraordinary resource on the history of the Port Washington sand mines, including archives of photographs, social histories, and a downloadable version of Elly Shodell's book *Particles of the Past.*]

Robertson, Barbara. "Spider-Man 3: Shifting Sand, Evil Goo." http://features.cgsociety.org/ story_custom.php?story_id=4091&page=2.

Roth, Siobhan. "History's X-ray." *National Geographic* 211 (April 2007): 27. [Geomagnetic imaging reveals the ancient city of Al Rawda.]

The Sand Castle. Directed by Co Hoedeman. 1977. www.nfb.ca/animation/objanim/en/films/ film.php?sort=title&id=12503.

"The Sandhog Project." www.sandhogproject.com.

U.S. Geological Survey. "Ground Water Atlas of the United States." http://capp.water.usgs .gov/gwa/gwa.html.

———. "Minerals Information." http://minerals.usgs.gov/minerals. [A comprehensive resource on all commercial minerals, in the United States and worldwide; see, in particular, silica, silicon, sand, aggregates, or any of the placer minerals.]

Wilson, James E. *Terroir: The Role of Geology, Climate, and Culture in the Making of French Wines.* London: Mitchell Beazley, 1998.

Youngquist, Walter. *GeoDestinies: The Inevitable Control of Earth Resources over Nations and Individuals.* Portland, OR: National Book Company, 1997. [A wide-ranging review of society's relationships with natural resources, with an emphasis on hydrocarbons.]

Zwingle, Erla. "Ogallala Aquifer: Wellspring of the High Plains." *National Geographic* 183 (March 1993): 80–109.

Chapter 10. Outward and Onward

Astrobiology Magazine. Home page. www.astrobio.net/news. [This NASA-hosted site is exactly what it says—an excellent and ongoing resource for accessible articles on wide-ranging topics of astrobiology.]

Battersby, Stephen. "Touchdown on Titan." *New Scientist* (January 14, 2006): 39–41.

Benison, Kathleen C. "A Martian Analog in Kansas: Comparing Martian Strata with Permian Acid Saline Lake Deposits." *Geology* 34 (May 2006): 385–88.

Boyd, Jade. "When Worlds Collide." *Sallyport: The Magazine of Rice University* (Fall 2003). www.rice.edu/sallyport/2003/fall/sallyport/worldscollide.html.

Chandler, David L. "Distant Shores: Mars Isn't the Place It Was Twelve Months Ago." *New Scientist* (January 15, 2005): 30–39.

Chavdarian, Gregory V., and Dawn Y. Sumner. "Cracks and Fins in Sulfate Sand: Evidence for Recent Mineral-Atmospheric Water Cycling in Meridiani Planum Outcrops?" *Geology* 34 (April 2006): 229–32.

Fookes, Peter G., and E. Mark Lee. "Climate Variation: A Simple Geological Perspective." *Geology Today* 23 (March–April 2007): 66–73.

"The Goldilocks Zone." *Astrobiology Magazine* (April 28, 2007). www.astrobio.net/news/ modules.php?op=modload&name=News&file=article&sid=2314.

Grotzinger, J. P., et al. "Sedimentary Textures Formed by Aqueous Processes, Erebus Crater, Meridiani Planum Mars." *Geology* 34 (December 2006): 1085–88.

———. "Stratigraphy and Sedimentology of a Dry to Wet Eolian Depositional System, Burns Formation, Meridiani Planum, Mars." *Earth and Planetary Science Letters* (November 30, 2005): 1–10.

Hansen, Kathryn. "Water Responsible for Martian Landscape?" *Geotimes* 52 (February 2007): 8–9.

Hoffman, Nick. "The Moon and Plate Tectonics: Why We Are Alone." *Space Daily* (July 11, 2001). www.spacedaily.com/news/life-01x1.html.

———. "Venus: What the Earth Would Have Been Like." *Space Daily* (July 11, 2001). www.spacedaily.com/news/life-01x2.html.

Horne, David J. "In the Land of the Meridian: Fieldwork on Mars." *Geology Today* 21 (January–February 2005): 27–32.

"Hot Air Could Halt Earth's Tectonics." *New Scientist* (April 19, 2008): 14. [On the future of plate tectonics; a brief summary of work by Adrian Lenardic and others.]

Irwin, Rossman P., III, Robert A. Craddock, and Alan D. Howard. "Interior Channels in Martian Valley Networks: Discharge and Runoff Production." *Geology* 33 (June 2005): 489–92.

Japan Aerospace Exploration Agency (JAXA). Home page. www.jaxa.jp/index_e.html.

"Kalahari Desert Sands an Important, Forgotten Storehouse of Carbon Dioxide." *Science Daily* (April 4, 2008). www.sciencedaily.com/releases/2008/04/080401200451.htm. [On bacteria in desert sand.]

Kasting, James. "Research: Habitable Zones around Stars and the Search for Extraterrestrial Life." www.geosc.psu.edu/~kasting/PersonalPage/ResInt2.htm.

Leeder, Mike. "Cybertectonic Earth and Gaia's Weak Hand: Sedimentary Geology, Sediment Cycling and the Earth System." *Journal of the Geological Society, London* 164 (2007): 277–96.

Muir, Hazel. "Second Rock from the Sun." *New Scientist* (June 5, 2004): 32.

Mullen, Leslie. "Extreme Animals." *Astrobiology Magazine* (September 1, 2002). www.astrobio.net/news/modules.php?op=modload&name=News&file=article&sid=261.

Näther, Daniela J., et al. "Flagella of *Pyrococcus furiosus:* Multifunctional Organelles, Made for Swimming, Adhesion to Various Surfaces, and Cell-Cell Contacts." *Journal of Bacteriology* (October 2006). http://jb.asm.org/cgi/content/full/188/19/6915.

National Aeronautics and Space Administration (NASA). "The Devils of Mars." *Science@ NASA* (July 14, 2005). http://science.nasa.gov/headlines/y2005/14jul_dustdevils.htm.

———. "NASA's Mars Exploration Program: Sand Dunes." http://mars.jpl.nasa.gov/gallery/sanddunes.

———. "The Sands of Mars." *Science@NASA* (January 31, 2005). http://science.nasa.gov/headlines/y2005/31jan_sandsofmars.htm.

The Planetary Society. Home page. www.planetary.org/home.

Schatz, Volker, et al. "Evidence for Indurated Sand Dunes in the Martian North Polar Region." *Journal of Geophysical Research* 111 (2006). www.icai.uni-stuttgart.de/~parteli/articles/Schatz_et_al_2006.pdf.

Sheehan, William. *The Planet Mars: A History of Observation and Discovery.* Tucson: University of Arizona Press, 1996. www.uapress.arizona.edu/onlinebks/mars/contents.

Shinbrot, Troy, Kiernan LaMarche, and Benjamin J. Glasser. "Triboelectrification and Razorbacks." http://sol.rutgers.edu/~lamarc/Razorbacks.

Ward, Peter, and Donald Brownlee. *The Life and Death of Planet Earth: How Science Can Predict the Ultimate Fate of Our World.* New York: Henry Holt, 2002; London: Piatkus, 2007.

———. *Rare Earth: Why Complex Life Is Uncommon in the Universe.* New York: Springer, 2000; New York: Copernicus Books, 2004. [This and *The Life and Death of Planet Earth* provide very readable, informative, and thought-provoking discussions of the past and future of our planet and the questions surrounding extraterrestrial life.]

Williams, Caroline, and Ted Nield. "Pangaea, the Comeback." *New Scientist* (October 20, 2007): 37–40. [On future continental scenarios.]

Epilogue

Boslough, Mark. "The Riddle of the Desert Glass" (September 15, 2006). www.cs.sandia.gov/newsnotes/DesertGlass_Sept15LabNews.pdf.

Cray Inc. "Cray Supercomputer at Sandia Helps Researchers Discover Origin of Mysterious Glass Found in King Tut's Tomb." (July 31, 2007). http://investors.cray.com/phoenix.zhtml?c=98390&p=irol-newsArticle&ID=1033862.

Giegengack, Robert, and James R. Underwood, Jr. "Origin of Libyan Desert Glass: Some Stratigraphic Considerations." In *Silica '96: Meeting on Libyan Desert Glass and Related Desert Events,* ed. Vincenzo de Michele. Milan: Pyramids Segrate, 1997.

Iredale, Will. "King Tut's Glass Beetle Came from Outer Space." *Sunday Times* (June 25, 2006). www.timesonline.co.uk/tol/news/uk/article679017.ece.

Wright, Giles. "The Riddle of the Sands." *New Scientist* (July 10, 1999): 42–45.

INDEX

DESIGNER: SANDY DROOKER

TEXT: 10/14 ADOBE GARAMOND

DISPLAY: AKZIDENZ GROTESK EXTENDED

COMPOSITOR: INTEGRATED COMPOSITION SYSTEMS

INDEXER: THÉRÈSE SHERE

PRINTER/BINDER: MAPLE-VAIL BOOK MANUFACTURING GROUP